D0209146

ENERGY *REVOLUTION*

ENERGY
REVOLUTION

THE PHYSICS *and the* PROMISE
of EFFICIENT TECHNOLOGY

Mara Prentiss

THE BELKNAP PRESS OF HARVARD UNIVERSITY PRESS

Cambridge, Massachusetts, and London, England

2015

Library of Congress Cataloging-in-Publication Data
Prentiss, Mara Goff.
    Energy revolution : the physics and the promise of efficient technology / Mara Prentiss.
        pages   cm
    Includes bibliographical references and index.
    ISBN 978-0-674-72502-7 (alk. paper)
  1. Power resources.   2. Energy consumption.   3. Energy conservation.   I. Title.
    TJ163.2.P735 2015
    333.793'2—dc23   2014020397

To my family

Contents

ENERGY *REVOLUTION*

INTRODUCTION

U.S. Energy Use—Past, Present, and Future

U.S. ENERGY SOURCES

A radical transformation of the energy economy from fossil fuel burning to renewable energy might seem highly improbable given human resistance to change, but an examination of historical energy use indicates that humanity has already undergone several energy consumption revolutions. Thus, another transformation is not only possible but probable. Predictions about the timing and nature of any such transition are much more difficult; however, it is certain that whatever happens must be in accord with the basic physical laws governing the universe.

Energy use has varied widely through time and from place to place at a given time. In the distant past, people extracted energy directly by burning wood, vegetable fat, or animal fat. Changes in energy consumption have been occurring for thousands of years; however, during the last 100 years the pace of change has accelerated enormously as energy use has skyrocketed, facilitated by dramatic changes in energy sources. These trends are depicted in Figure Intro.1 and Figure Intro.2, which illustrate the evolution of U.S. energy use. Figure Intro.1 shows total U.S. energy consumption, which has shown a steep increase over time, with the exception of three small periods of decrease. Those decreases occurred during the depression of the 1930s, the oil shock of the 1970s, and the global financial crash in 2008. U.S. energy has not yet fully recovered from the 2008 decrease.

In this book, I will show that renewable energy generated within the United States can meet the average total current energy needs, as well as the projected total energy needs for the next fifty years.[1] The question of whether renewable energy could provide all of the actual instantaneous energy needs of the United States is an open question that depends on how fluctuating renewable energy sources can be harnessed to provide power on demand. A revolutionary advance in large-scale energy storage would greatly ease the transition to a 100 percent renewable-energy economy; however, a combination of increases in energy efficiency due to widespread adoption of existing technologies and "smart grids" that pool energy supply and demand over large geographical areas may allow a renewable energy economy to flourish even without large-scale energy storage.

Given significant changes in energy use with time, it is difficult to use graphs showing energy use to extract how the balance of the U.S. energy portfolio has been changing, and in particular to determine whether the fractional contribution due to particular sources has changed significantly with time. For energy portfolio calculations, it is useful to consider the fractional energy contributions from different sources. This information is displayed in the lower of the two graphs in Figure Intro.2. That graph shows that energy use can be divided into three major eras on the basis of the dominant energy source: an age of wood, an age of coal, and an age of diversified consumption. Figure Intro.2 shows that during most of the age of wood, almost 100 percent of the energy used came from wood.

---

1. Renewable energy could meet all of the average U.S. energy demands by providing significantly less energy because burning fuel in cycling engines results in large energy losses that make the required fossil fuel energy much larger than the actual energy used, as highlighted in Figure Intro.10 and explained in Appendix A, which discusses why the Carnot efficiency limit constrains the energy efficiency that can be achieved in cyclic heat engines.

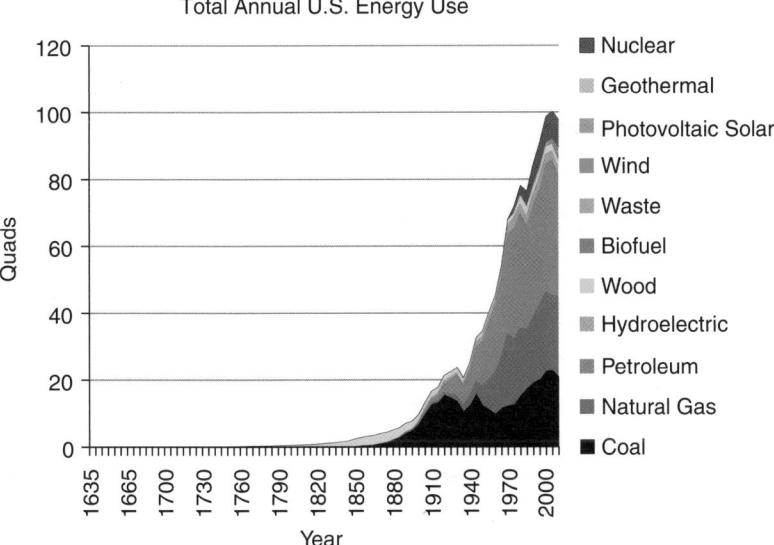

Figure Intro.1 Total U.S. energy consumption since 1635, showing contributions from different energy sources. Around 1900, U.S. total energy use began a steep increase that leveled off briefly in the 1970s and then grew again. Consumption dipped in 2008 and has not yet fully recovered. Source: Annual Energy Review 2011 United States Energy Information Administration, DOE/EIA-0384(2011), issued September 2012, tables 10.1, 10.3, and E1. Information available at www.eia.gov/totalenergy/data/annual/pdf/aer.pdf. Data from more recent years was obtained in topic 1.3, which is updated monthly. Information available at www.eia.gov/totalenergy/data/monthly/.

Similarly, during the age of coal, coal provided up to 80 percent of the energy. However, wood consumption decreased at the end of the age of wood, whereas coal consumption did not decrease at the end of coal. Figure Intro.2 also shows that coal use increased rapidly starting around 1850 and then remained fairly constant from around 1910 to 1970. The end of the age of coal was not the result of a decrease in coal use; rather, it was a consequence of large increases in the use of other sources. Petroleum consumption began a large increase in approximately 1910, followed by an increase in natural gas use that began in 1930. Finally, in the 1960s, nuclear power also began to make a significant contribution. In

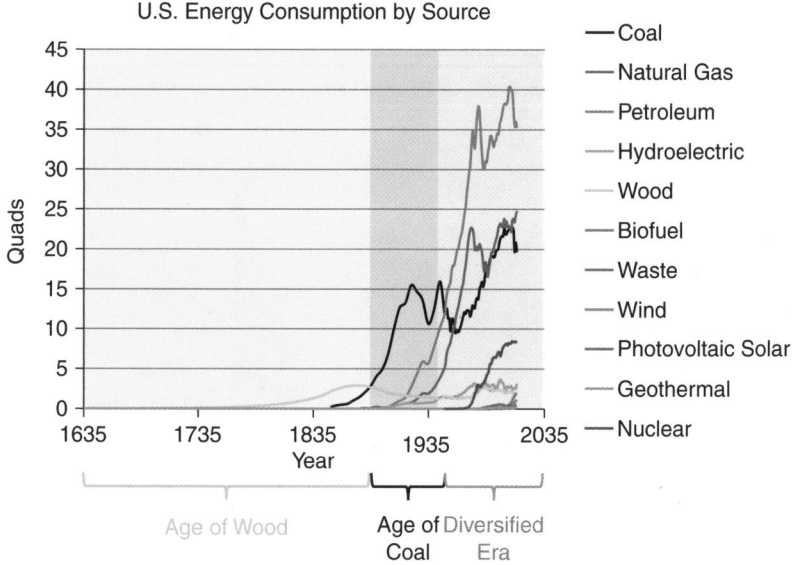

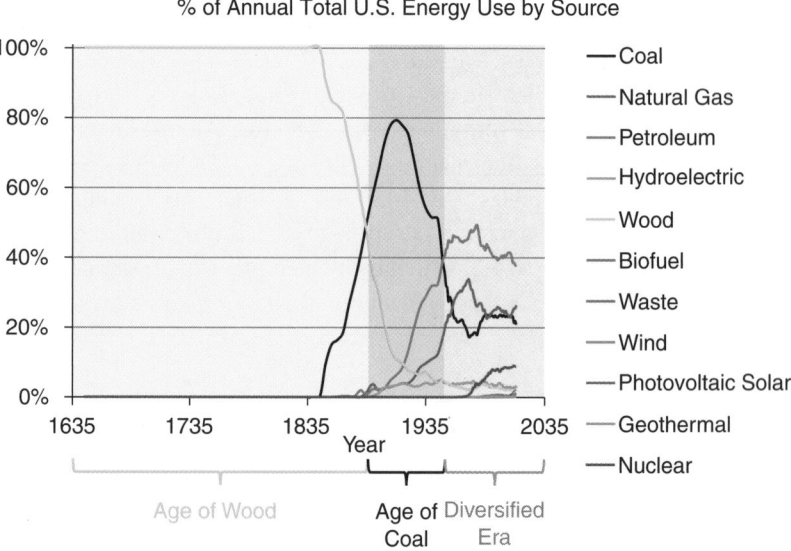

Figure Intro.2 U.S. primary energy consumption by source, 1635–2010. The upper panel shows the consumption in quads, and the lower panel shows the fractional consumption. In the age of wood, there was nothing else. By 1900, coal dominated. At present, oil makes the largest contribution, but coal and natural gas also make large contributions, with smaller contributions from nuclear and hydroelectric. Source: Annual Energy Review 2011 United States Energy Information Administration, DOE/EIA-0384(2011), issued September 2012, tables 10.1, 10.3, and E1. Information available at www.eia.gov/total energy/data/annual/pdf/aer.pdf. Data from more recent years was obtained from the information available in topic 1.3, which is updated. Information available at www.eia.gov/totalenergy/data/monthly/ monthly.

the diversified era, no single energy source contributes more than 50 percent of U.S. energy.

Even in this diversified era, coal consumption increased between 1980 and 2000, though coal use began decreasing again around 2005, as is clearly shown in Figure Intro.3, which illustrates consumption since 1949. Some of the decrease in the energy provided by coal is due to an overall energy use reduction resulting from the global financial crisis, but much of the recent dip is the result of electric power plants converting from burning coal to burning natural gas, as can be seen in the lower graph in Figure Intro.4. That graph clearly shows that the fractional contribution from coal has been shrinking, while the contribution due to natural gas has been rising. The conversion has significantly reduced carbon dioxide emissions in the United States. Interestingly, though the conversion from coal to natural gas has had a very favorable effect on emissions, it has been driven by market forces that do not consider the cost of emission. After 2008, U.S. natural gas prices substantially decreased as consequence of the increase in the supply of natural gas created by fracking.[2]

One might argue that the "age of wood" was an age of renewable energy, which was succeeded by two ages of nonrenewable energy. The energy contribution of wind has been increasing rapidly in both absolute and relative terms, even though the total contribution from wind remains so small that it is difficult to see in Figure Intro.2. Similarly, photovoltaic (PV) solar power has also increased very rapidly, but its absolute contribution remains so small that it is not visible in Figure Intro.2. The rapid increase in renewables might suggest that within the next twenty or so years renewable

---

2. In 2008, the average U.S. wellhead natural gas price was approximately $8 per Mcf,[2] but the price fell below $2.7 per Mcf in 2012, where Mcf represents a volume of natural gas. The M is an abbreviation for 1000 and cf stands for cubic feet. Thus Mcf represents 1000 cubic feet of natural gas. Similarly, MMcf represents $1000 \times 1000 = 1,000,000$ cf of natural gas. Based on a simple linear projection using the period from 2000 to 2005, the 2012 price would have been $18 per Mcf, whereas in fact it was $2 per Mcf; however, the projection from 2008 to 2010 almost exactly projects the 2012 value. Amusingly, continuing the projection to 2014 implies that natural gas would be free and that in 2015 producers would have to pay consumers to use it.

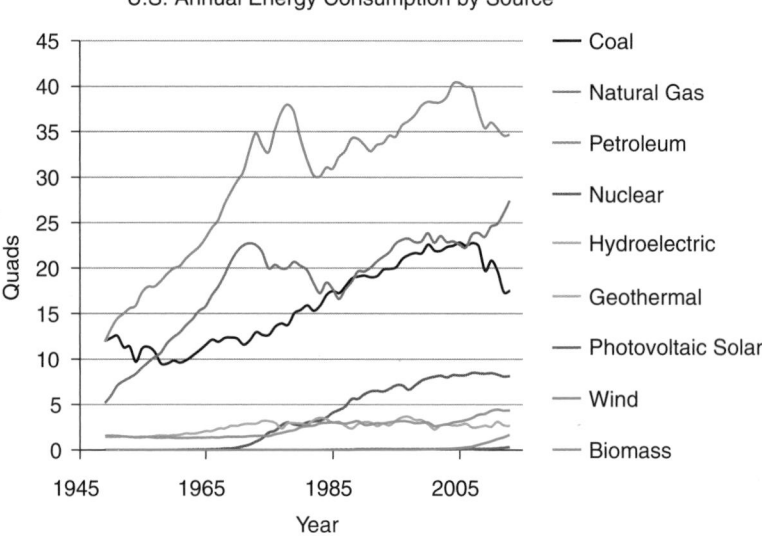

U.S. Annual Energy Consumption by Source

Figure Intro.3 Annual U.S. energy use by source in the diversified era from 1950 to 2011. Biomass includes wood and biofuel. Hydroelectric power has been constant since around 1970. Nuclear increased steadily until around 2000 and began to show a distinct decrease in 2011 following the closing of four nuclear plants. Renewable sources now provide more energy than nuclear sources do, with renewables rapidly increasing and nuclear stagnating. Coal has been falling since around 2005, partly as a result of the increase in natural gas occurring at the same time largely because the development of fracking as an extraction technology has increased supply and decreased prices. The sum has maintained close to a constant level since around 1995. Source: Annual Energy Review 2011 United States Energy Information Administration, DOE/EIA-0384(2011), issued September 2012, tables 10.1, 10.3, and E1. Information available at www.eia.gov/totalenergy/data/annual/pdf/aer.pdf. Data from more recent years was obtained from topic 1.3, which is updated monthly. Information available at www.eia.gov/totalenergy/data/monthly/.

energies could dominate the U.S. economy; however, as recent stock market results have shown, trends do not necessarily continue, or as Yogi Berra said, "It is difficult to make predictions, especially about the future."

Though long-term historical data are interesting, it is the most recent data that may be most useful in making projections about the near future.

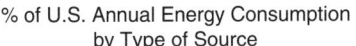

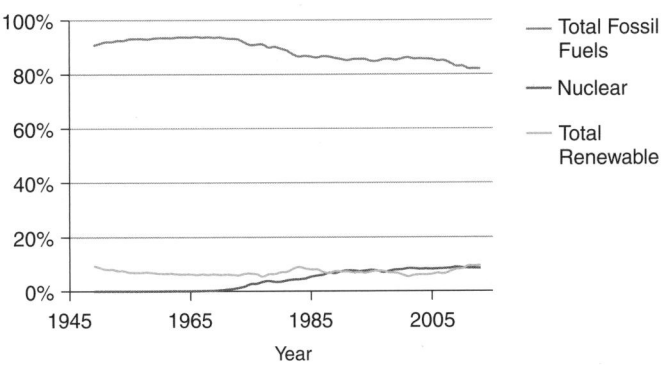

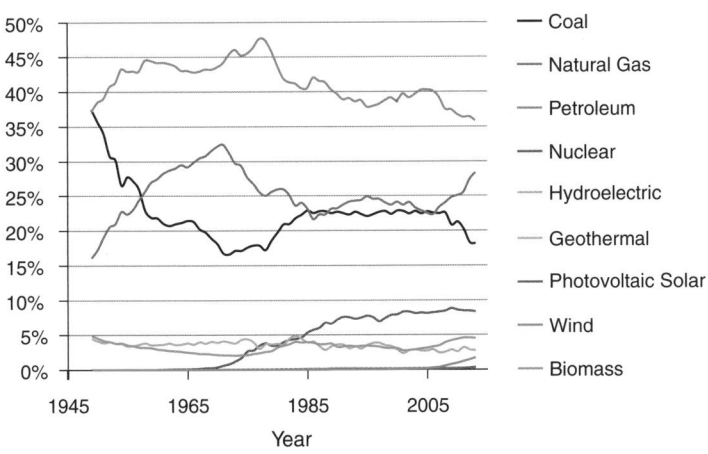

Figure Intro.4 Percentage of U.S. annual energy consumption by source category (upper) and source (lower). The charts show that the contribution of fossil fuels remains dominant at more than 80 percent. Given that total energy use has been increasing, a constant energy contribution represents a decreasing fraction of the total energy consumption. The charts show that the fractional contribution of fossil fuels decreased from 1975 to 1985, followed by a stable period up to around 2005. The stable period was followed by a decline in 2008, which has been followed by an overall stabilization. Nuclear energy increased fairly rapidly from 1960 to 1990, followed by a very slow increase, leading to a more or less flat contribution since 2005. In 2013, the fractional contribution of renewables was just slightly higher than the contribution in 1950, but the contribution of different renewables has changed significantly. In topic 1.3 at www.eia.gov/totalenergy/data/monthly/, which is updated monthly.

Thus, Figure Intro.3 and Figure Intro.4 consider U.S. energy consumption only in the diversified era, where no source provided more than 50 percent of the energy.

The upper graph in Figure Intro.4 indicates that the overall fractional contribution of renewable energy has remained stable since around 1949, which seems to belie all of the hype about the rise of renewable energy; however, this stability masks a shift from hydroelectric power to biofuels, as well as a recent explosion in the energy provided by wind and solar PV, as will be discussed in Chapter 2.

U.S. ENERGY USES

When considering questions involving energy, it is important to know both where the energy comes from and how the energy is used, particularly since in this book I will argue that reductions in use may make a significant contribution to the transition to a renewable energy economy. In 2012, the United States used approximately 100 quads of energy,[3] and the entire world used approximately 450 quads. "Quad" is short for a quadrillion British thermal units (BTUs), which is not a standard international measure; however, quads are frequently used in national and international discussions of energy because the resulting numbers are in the hundreds. Though energy consumption on a national scale is often measured in quads, the standard international (SI) unit of energy is a joule, where one quad is approximately equal to 10^{18} joules, which is referred to as an exajoule. Table 2 provides a summary of the prefixes that are used to indicate powers of ten in SI units. Conversions between different units of energy are shown Table 3, including the approximate equivalence of 1 quad and 1 exajoule.

The pie chart in Figure Intro.5 shows the fraction of energy that the United States obtained from different primary sources in 2012. Because

3. According to the Energy Information Agency Report DOE/EIA-0383ER(2014), the U.S. total energy consumption in 2012 was 95 quadrillion BTU.

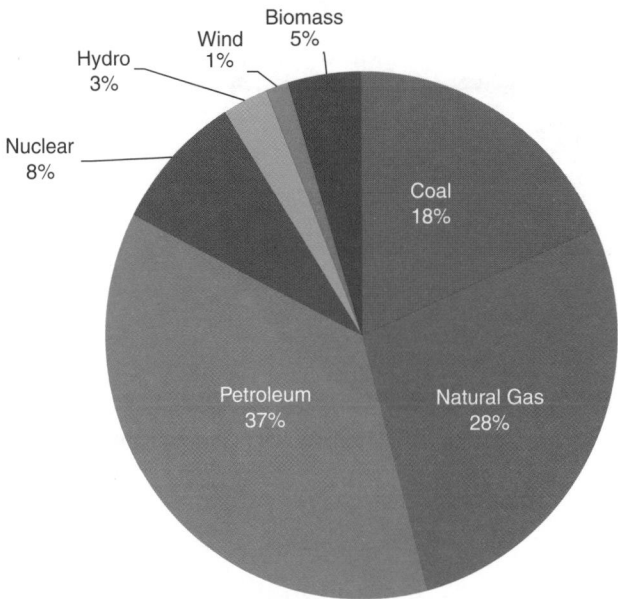

U.S. Primary Total Energy Sources 2012

Biomass 5%
Wind 1%
Hydro 3%
Nuclear 8%
Coal 18%
Petroleum 37%
Natural Gas 28%

Figure Intro.5 Primary sources of U.S. energy in 2012 in units of quadrillion British thermal units. From U.S. Energy Information Administration in topic 1.3, available at www.eia.gov/totalenergy/data/monthly/#summary.

the total number of quads used is almost 100, the number of quads and the percentages are about the same. The chart shows that together all of the renewable energy combined contributes only slightly more than nuclear power and that petroleum contributes the largest fraction of energy use.

In order to understand the flow of energy through the U.S. economy, understanding end uses is as important as understanding sources. Figure Intro.6 divides energy use into sectors. The upper graph shows that electrical power, using 40 quads, dominates total energy consumption, with transportation coming in second at 27 quads. The lower graph reveals large differences in how primary sources are used. All of the nuclear energy and the vast majority of the coal energy go into generating

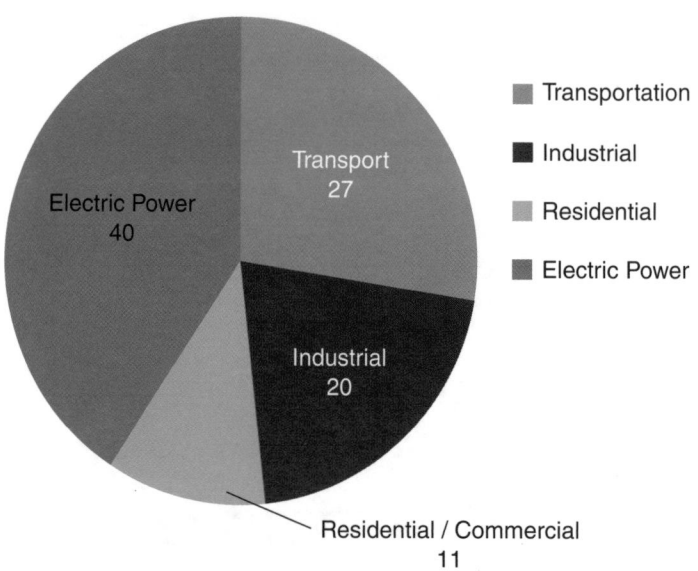

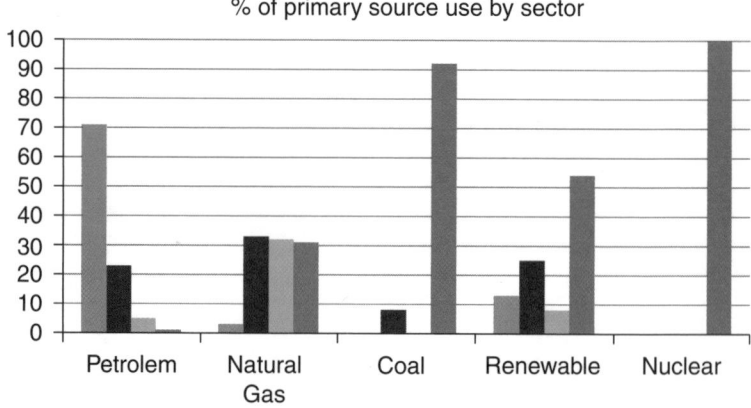

Figure Intro.6 Energy use by sector. The upper graph shows the number of quads used by each sector. It shows that energy use is dominated by electric power generation and transportation. Since the total U.S. energy consumption in 2011 was 98 quads, the consumption in quads is approximately the same as the percentage of consumption, and percentage of U.S. primary energy sources used by sectors of the U.S. economy in 2011. From U.S. Energy Information Administration in topic 2.1, available at www.eia.gov/totalenergy /data/monthly/#summary.

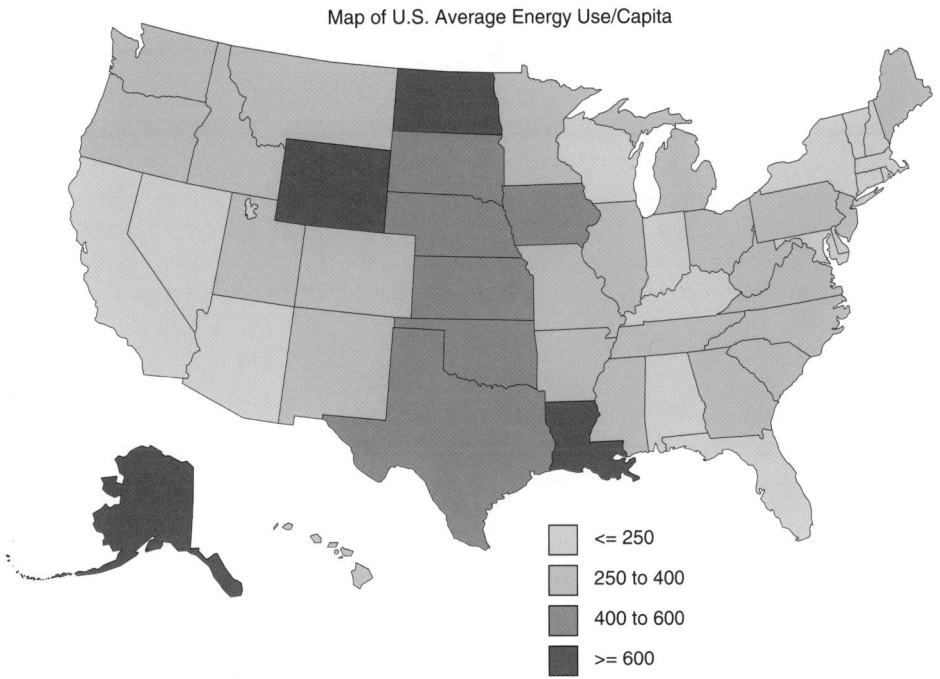

Figure Intro.7 U.S. total energy consumption per capita by state in billions of British thermal units. The map was made by the U.S. Energy Information Administration using data from 2011. A current version of the map and data are available at www.eia.gov/state /rankings/?sid=US.

electricity, whereas petroleum contributes almost nothing to electrical generation, even though it is the largest overall primary source. Transportation dominates the petroleum use, with most of the remaining petrolum use confined to industry. Natural gas contributes almost nothing to transportation; however, its contributions are approximately evenly distributed across the other use categories. Finally, approximately half of the renewable energy goes into generating electricity, with industry using another quarer, and the remainer divided between transportation and industry.

Average values for the United States can mask important regional differences. Figure Intro.7 is a map illustrating the average annual U.S. energy

Table 1.   Table of the average energy annual consumption per capita, by state, in billions of British thermal units, ordered from lowest to highest.

State	Consumption	State	Consumption	State	Consumption
RI	175	CO	289	AR	380
NY	185	DC	291	WV	390
CT	207	PA	292	MS	391
HI	208	VA	295	MT	398
CA	209	DE	299	AL	402
MA	211	WA	305	KS	405
AZ	221	GA	306	OK	421
FL	221	IL	309	KY	438
NH	222	ME	311	IN	440
NV	233	MO	313	SD	464
VT	238	WI	313	NE	473
MD	244	NM	331	TX	476
OR	262	ID	332	IA	494
NC	267	OH	332	ND	768
NJ	276	TN	344	AK	881
UT	283	SC	345	LA	886
MI	284	MN	349	WY	975

Source: U.S. Energy Information Administration 2011 data.

use per capita; the corresponding numerical values are provided in Table 1. In general, though cities use more power than rural areas, the energy use per capita in densely populated urban areas is much lower than in suburbs or rural areas.[4] The lower consumption in urban areas probably results from a number of factors, including (1) smaller living spaces in large buildings;[5] (2) the use of public transportation and walking; and (3) smaller required driving distances. When considering opportunities to reduce consumption, it is important to keep such regional differences in mind.

4. See www.hks.harvard.edu/var/ezp_site/storage/fckeditor/file/pdfs/centers -programs/centers/taubman/policybriefs/greencities_final.pdf.
5. The surface-to-volume ratio in apartment buildings is lower than in single-family houses, reducing energy waste associated with heat loss.

OPPORTUNITIES TO REDUCE
ENERGY CONSUMPTION

In this section, I provide a few examples of how changes in technology have produced significant reductions in energy use, and later in the book I will explore additional opportunities for reducing energy consumption, with an emphasis on changes that reduce fossil fuel use. As Figure Intro.1 shows, U.S. energy consumption has been increasing rapidly, with significant increase in wood use that began when the United States was still a British colony. The increase in energy has been due both to increase in population and to increases in energy use per person, where the latter has historically also been linked with increases in gross domestic product (GDP). It has long been feared that attempts to decrease energy consumption would have a direct adverse effect on quality of life as well as an indirect adverse effect resulting from reductions in economic growth. In this book I will argue that making an economically feasible transition to a renewable energy economy may depend as much or more on our ability to decrease energy use as it does on our ability to increase renewable energy harvesting. Suggestions that people decrease energy use have a very negative history in the United States; however, recent U.S. data suggest that increases in energy efficiency and changes in technology mean that decreases in energy use may not require sacrifices. Furthermore, some energy-saving changes have been accompanied by lifestyle improvements.

As Figure Intro.1 shows, total energy use has increased by around 100 percent since 1960, with a slight decrease since 2008; however, Figure Intro.8 shows that energy use per capita has remained approximately the same since 1960, albeit with fluctuations over the decades. Energy use per capita fell below the 1970 level in about 2000 and has now fallen to approximately the 1965 level. The decrease in energy use per capita illustrated in Figure Intro.8 demonstrates that the energy-use increase since 1960 that is shown in Figure Intro.1 is predominantly due to the approximately 75 percent increase in population that has occurred

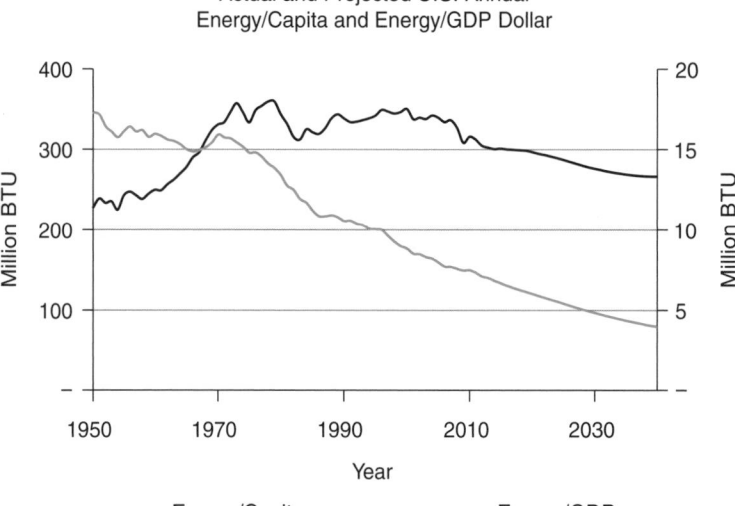

Figure Intro.8 U.S. annual energy consumption per capita and per GDP dollar showing both the actual and projected data from the U.S. Energy Information Administration (EIA), indicating that the energy per GDP has shown a steady and substantial decrease, so that the 2011 value is just slightly more than one-half the value in 1980. In contrast, the energy per capita has fluctuated, with 2011 values around 90 percent of the values in 1980. The EIA projections through 2040 indicate a continued modest decrease in energy per capita and an almost 50 percent reduction in energy per GDP dollar. From U.S. Energy Information Administration in topic 1.7, available at www.eia.gov/totalenergy /data/monthly/#summary.

since 1960. Better still, the amount of energy required to create a dollar of GDP has been decreasing rapidly since around 1970. The U.S. Energy In- formation Administration (EIA) projects a continued decrease in energy use per GDP$, leading to another reduction by a factor of two by 2040.

The black curve in Figure Intro.8 is remarkable: despite the explosion in electronic devices, the supersizing of U.S. houses, and the enormous increase in miles driven, energy use per capita has remained the same since approximately 1965. This constant level of energy use per capita is

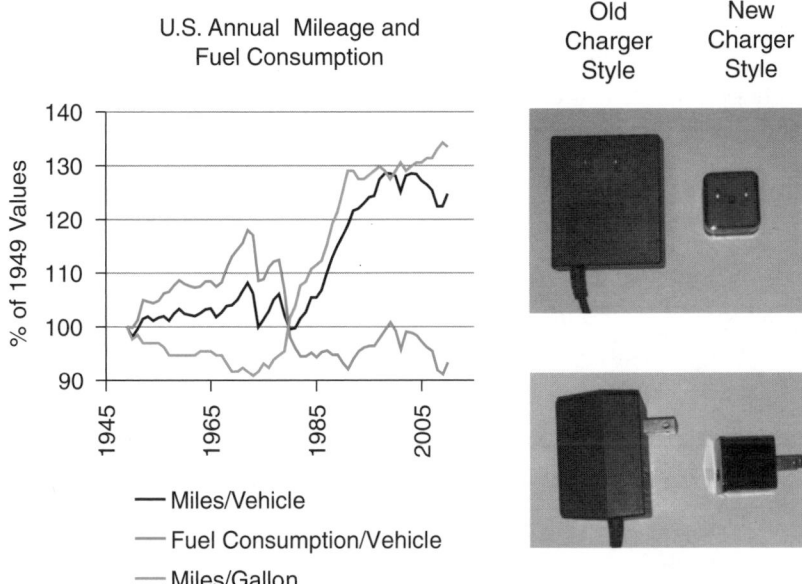

Figure Intro.9 How technical changes reduce energy use without sacrificing lifestyle. At left, U.S. transportation data since 1949 show total annual miles driven per vehicle as well as the fuel consumed to drive those miles. The graph shows that though the number of miles driven has increased by more than one-third, the gasoline used has decreased by approximately 10 percent as a result of increases in the number of miles per gallon. The information is available in topic 1.8 at www.eia.gov/totalenergy/data/monthly/#summary. The photo at right shows old analog chargers as well as new digital chargers. The change in technology has reduced both volume and weight by more than a factor of ten. The technology shift has also improved energy efficiency by a factor of more than three and makes it possible for the same charger to be used anywhere in the world.

the result of several factors. One is the increased energy efficiency of devices, such as cars and appliances. Figure Intro.9 shows that the annual number of miles driven per vehicle has increased by more than a third since 1949; however, increases in fuel efficiency have been so large that the present U.S. fuel consumption is almost 10 percent lower than it was in 1949.[6] Furthermore, this increased energy efficiency has been coupled with dramatic decreases in emissions, as I will discuss in detail in Chapter 10. This reduction in energy use and decrease in undesired emissions has not been achieved at the expense of quality of life: very few people would argue that driving (or riding in) is less satisfying in a modern car than in a 1949-era car, even without including the advantages offered by cupholders, MP3-based stereo sound systems, and GPS navigation.

Decreases in energy efficiency have come not only from incremental technical improvements but also come from revolutions in technology. The switch from incandescent lightbulbs to LED lightbulbs involves a complete change in how the light is produced, as I will discuss in Chapter 8. The shift from incandescents to LEDs can reduce the energy required by lighting by a factor of four because most of the energy used by incandescents is released as waste heat. The waste heat from those bulbs was large enough to bake cakes in the "easy bake ovens" that were popular in the 1960s. The large energy savings that result from switching lighting sources has not resulted in any lifestyle sacrifice since the new generation of LED lightbulbs provides light that is very similar to that produced by incandescents. In contrast, the light from compact fluorescents carried a purple-blue tinge that many people found quite unpleasant, which was one factor that limited the willingness of consumers to trade incandescent bulbs for compact fluorescent bulbs.

6. In 2010, the average vehicle was driven 12,000 miles, and the average fuel efficiency was 17.5 miles per gallon, so the average vehicle consumed 685 gallons in one year. At a gasoline price of $3.50 per gallon, the total annual gasoline cost was $2,400. In 1949, the corresponding numbers were 9,600 miles driven at an average fuel efficiency of 13 miles a gallon, resulting in an average fuel consumption of 733 gallons. This represent an inflation adjusted annual fuel cost of ~$1,800 given that the inflation adjusted price of gasoline is approximately $2.50/gallon (inflationdata.com/Inflation/Inflation_Rate/Gasoline_Inflation.asp). Thus, the average annual gas expenditure has increased by approximately 30 percent despite a decrease in use of approximately 10 percent.

Though it is good to preserve lifestyle, it is better still to improve it, and technological revolutions can provide both decreases in energy use and increased pleasure for the user. A wonderful recent example is the chargers that plug into the wall to provide power for portable devices such as cell phones. Before 2005, the chargers contained coils of wire. One such device is shown in Figure Intro.9, where it is compared with its digital replacement. The old wire coil charges were large and heavy and could not work in both Europe and the United States because European and American outlets deliver different amounts of voltage. More recent devices are digital. As illustrated in Figure Intro.9, these new devices are approximately ten times smaller and more than ten times lighter, and they will work with any voltage source in the world.

So far, I have been considering only those dramatic energy reductions that have already occurred or are occurring. It is also important to know whether there are more technical opportunities for reducing energy consumption. The pie chart in Figure Intro.10 shows the fraction of energy wasted in the United States in 2012. The graph is arresting. It shows that a substantial majority of the energy we obtain from primary sources is wasted. This graph would seem to present an enormous opportunity. If only we could stop wasting so much energy, many of our energy problems would be solved. In this book, I will consider how much of that "wasted" energy really represents an opportunity for improvement and how much is a consequence of basic scientific laws that cannot be altered.

One such significant law is the second law of thermodynamics, which requires that energy must be wasted in any process that burns fuel to drive a cyclic heat engine.[7] This law is particularly important because fossil fuel–burning power plants and internal combustion engines are cyclic heat engines, though gas furnaces and jet engines are not. The

7. The Carnot limit states that the maximum possible efficiency is given by $1 - T_{cold}/T_{hot} \sim 0.5$, where T_{cold} is around 25°C, room temperature is around 300 degrees Kelvin, and T_{hot} is the temperature inside the hot engine, approximately 350°C = 650 degrees Kelvin. Carnot efficiency is discussed in detail in Appendix A.

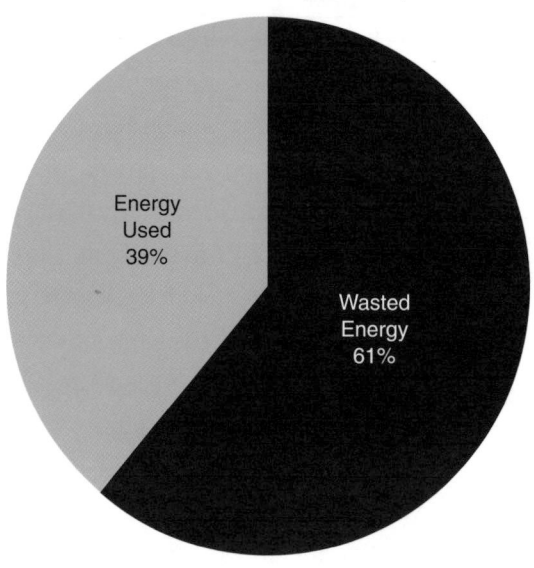

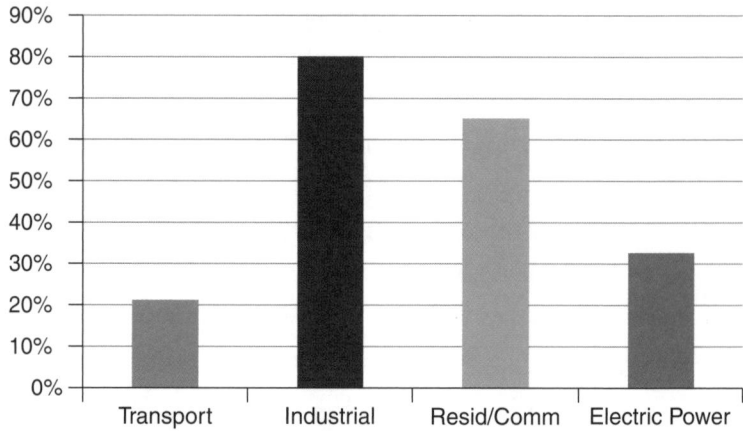

Figure Intro.10 Fraction of U.S. primary energy that was used and wasted in 2012. The upper graph shows that almost two-thirds of the total energy consumed in the United States is wasted. The lower graph displays energy efficiency by sector, which shows the poor efficiency for transport and electric power generation. Data from Lawrence Livermore National Laboratory energy flow chart, www.llnl.gov/news/newsreleases/2013/Jul/images/28228_flowchart highres.png. Also available as www.eia.gov/totalenergy/data/monthly/pdf/flow /total_energy.pdf.

lower graph in Figure Intro.10 shows that industry is almost 80 percent efficient and that the residential-commercial sector is approximately 65 percent efficient. In contrast, electric power is only around 40 percent efficient, and transport is only around 20 percent efficient. The second law of thermodynamics accounts for much of the waste in transport and electrical power generation, where cyclic heat engines play a large role. Thus, an important question is how much of that waste could be improved by engineering and how much of it is fundamental and unchangeable.

For typical gasoline engines and electric power–generating turbines, the second law of thermodynamics gives theoretical maximum efficiency of around 55 percent. For electrical generators, typical efficiencies are approximately 42 percent for natural gas–burning plants but only around 33 percent for coal burning plants.[8] It is noteworthy that the efficiency for coal was approximately stable from 2002 to 2012, whereas the efficiency for natural gas has increased steadily from less than 36 percent in 2002 to more than 42 percent in 2012, highlighting how rapidly improvements in energy efficiency can occur; however, that improvement also highlights how close we are to the theoretical efficiency limit and how little additional advantage we might obtain from improved engineering.

The energy efficiency of cars is much worse than for electric power plants. Automobiles driven in the city have energy efficiencies of about 15 percent,[9] owing to additional energy losses not associated with the second law of thermodynamics. In other words, greater than 6 joules' worth of energy from a gasoline is required to provide 1 joule of energy that moves. Though gasoline and diesel engines waste the majority of the energy in the petroleum that powers them, commercial electric motors can be more than 99 percent energy efficient. Furthermore, electric motors do not themselves emit smoke, soot, or chemical emissions. Thus, switching from petroleum-powered vehicles to electric vehicles could improve

8. See www.eia.gov/electricity/annual/html/epa_08_01.html.
9. See Figure 8.3.

energy efficiency. If the electricity was generated from renewable sources that don't require combustion, then the switch to renewable energy would greatly improve energy efficiency and also virtually eliminate emissions. Even if the electricity were produced by a natural gas–burning plant, emissions of some pollutants would be substantially reduced;[10] however, if the electricity is generated by a coal-burning plant, it is not clear that emissions would be reduced or that energy efficiency would increase. Given that at present the United States generates approximately twice as much electricity from coal as from natural gas,[11] it is not clear that switching to electric cars today would be advantageous. Of course, in a future where renewable energy or even natural gas is the dominant source of electric power, the conversion to electric cars would be extremely favorable.

The "wasted" energy in fossil fuel–burning engines or turbines also has very important implications when considering the feasibility of switching to renewable energy. In order to replace fossil fuels, renewable sources need only to meet the needs of energy consumers. For example, only around 1 joule of electrical energy from a renewable source is required to replace the more than 6 joules of gasoline energy required to propel a car through the city.

Many energy books make very optimistic assumptions. To avoid unrealistic discussions of the feasibility of renewable energy, I will present feasibility in three different ways: (1) renewable energy as a 100 percent replacement of all consumed fuels, including energy wasted in heat engines, which is the absolute possible upper limit and quite unrealistic; (2) renewable energy as a replacement of energy actually used, which is less than 50 percent of the total energy and fairly realistic; (3) renewable energy as a replacement of only the electrical energy delivered to consumers without replacing any of the energy now not provided by electrical

10. See Figure 10.3.
11. See Figure 2.6.

Table 2. Numerical significance of SI prefixes. For example, "kilo-" represents 1,000. Thus, 1 kilometer is 1,000 meters and 1 kilogram is 1,000 grams.

yotta- (Y-)	10^{24}	1 septillion
zetta- (Z-)	10^{21}	1 sextillion
exa- (E-)	10^{18}	1 quintillion
peta- (P-)	10^{15}	1 quadrillion
tera- (T-)	10^{12}	1 trillion
giga- (G-)	10^{9}	1 billion
mega- (M-)	10^{6}	1 million
kilo- (k-)	10^{3}	1 thousand
hecto- (h-)	10^{2}	1 hundred
deka- (da-)	10	1 ten
deci- (d-)	10^{-1}	1 tenth
centi- (c-)	10^{-2}	1 hundredth
milli- (m-)	10^{-3}	1 thousandth
micro- (μ-)	10^{-6}	1 millionth
nano- (n-)	10^{-9}	1 billionth
pico- (p-)	10^{-12}	1 trillionth
femto- (f-)	10^{-15}	1 quadrillionth
atto- (a-)	10^{-18}	1 quintillionth
zepto- (z-)	10^{-21}	1 sextillionth
yocto- (y-)	10^{-24}	1 septillionth

sources.[12] In all three cases, for renewables I always include a factor that relates the rated capacity values, also known as nameplate power values, for wind and solar systems to the actual average power delivered based on existing commercial units.

There is a second very important type of "wasted" energy that is not considered in Figure Intro.10. The gasoline burned while driving around searching for an urban parking place[13] or sitting still in a traffic jam

12. Transmission losses for renewable and nonrenewable energy are similar for any particular power line, however, renewables often require longer power lines. In addition, offshore windfarms usually require that power be transmitted under the ocean, where losses are larger. For some cases, like long-distance power transmission under the ocean, the assumption may be a bit optimistic. This issue will be considered in Chapter 7.

13. The energy wasted looking for parking is estimated to be 1 million barrels of oil a day. See www.csmonitor.com/Environment/Energy-Voices/2013/1030/How-smart -parking-could-save-a-million-barrels-of-oil-every-day.

Table 3. Relationship of various common energy-measuring units to the SI standard joule.

Name of unit	Symbol	Definition	Origin or common use	Approximate equivalent number of joules
SI units				
Joule	J		Work done to move an object 1 m pushing against a force of 1 N (1 N · m, which has SI units of 1 kg m²/s²), also 1 watt of power consumed for 1 second (1 W · s)	1
Exajoule				10^{18} J
Other units				
British thermal unit	Btu		Heat required to raise the temperature of 1 pound of water by 1°F at a pressure of 1 atm	1,000 J
Quad	quad	10^{15} Btu		$1. \times 10^{18}$ J or ~1 exajoule
Calorie	cal	4.1868 J	Energy required to raise 1 g of water by 1°C at a pressure of 1 atm	4.2 J
Food calorie (really kilocal)	Cal	1,000 calories	Used on food labels to indicate energy content	4,200 J
Electronvolt	eV	e × 1 V	Work done to move an electron between two points with a potential difference of 1 V	1.6×10^{-19} J
Mega electron volts	MeV	1,000,000 eV	Work done to move an electron between two points with a potential difference of 1,000,000 V	
Kilowatt-hour	kW · h		Used to bill electricity equal to the energy used when consuming 1 kW of power for 1 hour	3.6×10^{6} J
Kilowatt-year	kW · year		Approximate annual U.S. household electricity use	3.1×10^{10} J
Megawatt-year			Useful windpower unit	3.1×10^{13} J
Therm		~100,000 Btu	Used to bill natural gas	1×10^{8} J = 100 MJ

during a commute to work does not appear in this chart, but it is energy that is consumed with minimal benefit to the consumer. Other examples are "vampire" chargers that are left plugged in but not charging any device, lights left on in unoccupied rooms, and poorly insulated buildings. Such energy waste causes us to consume energy that we do not really use. In this book, I will consider how advances in electronics enable "smart" energy use, which could greatly reduce such losses. Though scientists and engineers tend to emphasize changes that depend on technical features of devices, energy use choices made by individual people may decrease the negative consequences of energy use much more than the action of scientists and engineers.

FOUNDATIONS OF A RENEWABLE FUTURE I

1

OVERVIEW OF RENEWABLE ENERGY

Though nonrenewables dominate the energy economy, they cannot continue to dominate forever because their supply on earth is finite. Discoveries such as fracking allow people to use fossil fuel from previously inaccessible reservoirs, which may extend fossil fuel use well into the next century; however, the total quantity of fossil fuel in the earth is finite, so eventually we will run out.[1]

One can describe the amount of fossil fuel in the earth in terms of the total energy stored in those fossil fuels. There are many units of energy that are used for different purposes, as is discussed in more detail in online Appendix 3 at http://thedata.harvard.edu/dvn/dv/HUP. The standard measurement system for scientific research is called standard international units, or SI units. In SI units, time is measured in seconds, distance is measured in meters, and mass is measured in kilograms, as discussed in detail in online Appendix 2. Most important for the context of this book, in SI units, energy is measured in joules. Thus, the total amount of stored

---

1. Fossil fuels are not truly finite: they are renewed on geological timescales, but the rate is infinitesimal in comparison with our energy use rate, so it is reasonable to assume that there will be no increase in the amount of fossil fuel during the next several thousand years.

fossil fuel energy can be expressed in terms of joules.[2] The problem with nonrenewable energy is that once we spend all of those joules, we are out of energy; therefore, continued development for expanding our use of renewable energy is necessary.

Figure 1.1 is a graph of U.S. non–fossil fuel energy production from 1949 to 2013, which shows that in 2012 the energy contribution of renewables surpassed the energy contribution of nuclear power. Lumping all nonrenewables together masks important changes in the contributions of different renewable energy sources, as illustrated by the upper right graph in Figure 1.1. The graph shows that hydroelectric power has been effectively constant since the mid-1970s. In Chapter 3, I will argue that U.S. exploitation of hydroelectric power is already approximately maximized, so the absolute energy contribution of hydroelectric power is likely to remain constant. Biomass was flat from 1949 to the mid-1970s, when there was a big increase after the oil shock, followed by a second increase beginning in around 2000 and continuing into the present. The biofuels increase owes a good deal to regulation that requires that gasoline contain ethanol. Windpower made a negligible contribution until around 2005, after which it began rising rapidly. I will argue that both biomass and wind power can continue to show substantial increases in the future. Geothermal and photovoltaic solar (PV) each continue to make such a small contribution that they are not readily visible in a graph showing the total renewable energy, so Figure 1.2 shows only the contributions of photovoltaic solar (PV), wind, geothermal, and ethanol in order to highlight the rapid recent increases in PV and wind.

Despite the dramatic increases in renewable energy that are shown in Figure 1.2, the EIA does not think that the U.S. energy use balance is going to change significantly during the next thirty years. The EIA projects that our energy balance will remain almost exactly the same as it is

---

2. In the Introduction, we measured energy in quad = 1 quadrillion BTU, which conveniently is ~10^{15} joules = 1 exajoule.

now and that overall energy use will increase only slightly. In the rest of this book, I will demonstrate that the very limited renewable energy growth the EIA projects is not the result of limitations on the availability of renewable energy.

One goal of this book is to provide people with information that allows them to make decisions about energy use. It is entirely possible that those decisions will lead to precisely the future predicted by the EIA, but it is also possible that our individual and collective decisions will result in a dramatically different future in which renewables play a much larger role and total energy use may actually decrease over time without lifestyle sacrifices.

THE SUN IS ENOUGH

In contrast to fossil fuels, "renewable" energy resources are effectively inexhaustible because they are constantly replenished by the sunlight shining on the earth. In this book, I will assume that the sunlight will continue to fall on the earth forever, though astronomers would beg to differ;[3] therefore, given that the sun will continue to shine "forever," the first fundamental question governing the feasibility of converting entirely to a renewable energy economy is whether the sun delivers energy faster than we consume it. To answer this question, we need to know how fast the sun delivers energy and how fast we consume it. In SI units, the rate at which energy is delivered or consumed is measured in joules per second, where one joule per second is called a watt. Since power is defined as the change in energy over time, in SI units the watt is a measure of the amount of power produced or consumed. This is why lightbulbs are rated in watts. The watt rating indicates how fast the bulb must consume energy in order to emit light. The energy used by the lightbulb is not listed

3. The sun is approximately 4.6 billion years old, and it is expected to shine for approximately another 5 billion years; however, between now and then the sun will actually heat up. The oceans will boil before the sun stops shining.

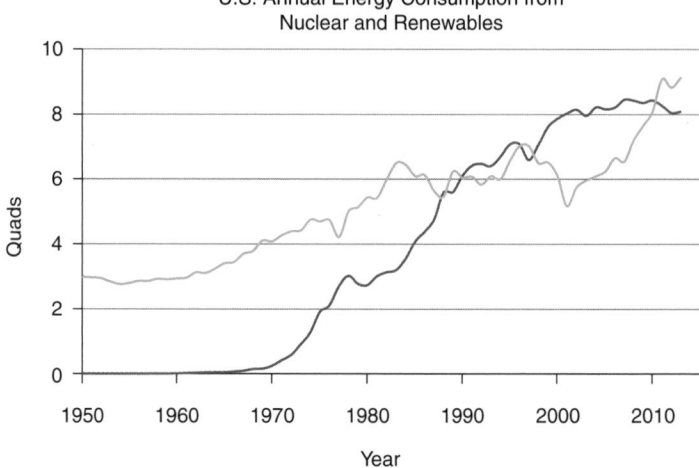

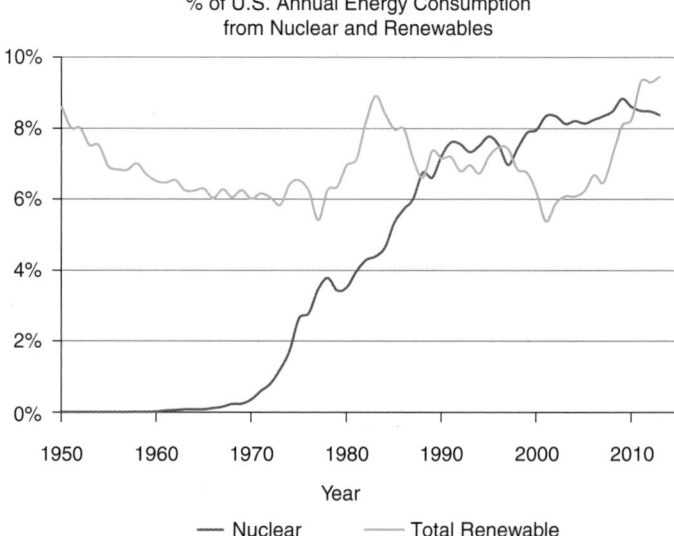

Figure 1.1 U.S. annual non–fossil fuel and renewable energy consumption by source. The upper left graph shows the consumption in quads, and the lower left graph shows the percentage of total consumption. The graphs show that nuclear and total renewable energy made comparable contributions from 1985 to 1995, when renewables showed a sharp decrease and nuclear showed a slight increase. The stagnation in nuclear since around 2000 coupled with the marked increase in total renewables made the contribution of total renewables exceed the contribution from nuclear power plants in 2012. The upper right graph shows the consumption in quads, and the lower right graph shows the percentage of total consumption. Hydroelectric has shown a steady decline because

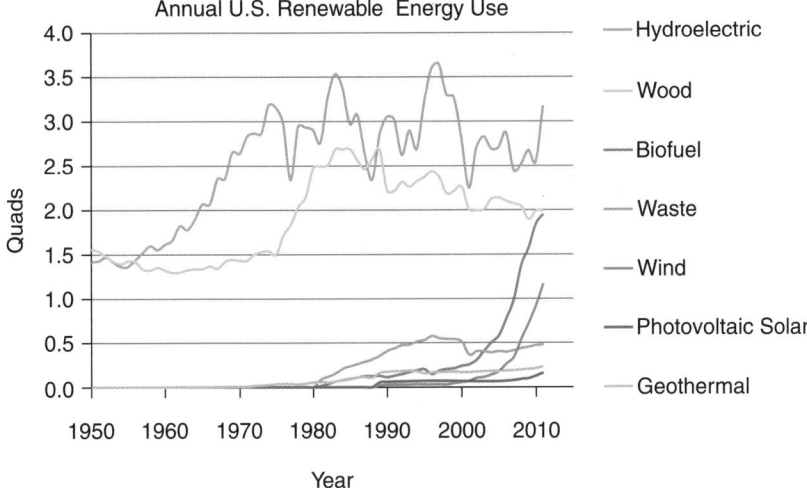

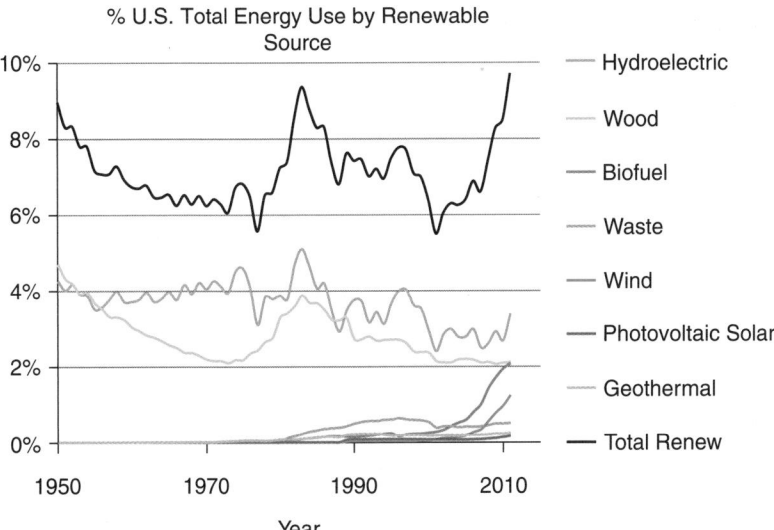

its constant absolute contribution represents a decreasing percentage of the total energy used. In these graphs, biomass is broken down into wood and biofuel, which allows one to see that wood has been decreasing since around 1985, whereas biofuels have been rapidly increasing since around 2000. Half of the biofuel contribution is from ethanol in gasoline as a result of regulatory actions encouraging ethanol as a motor fuel, as shown in Figure 1.2. Wind power has been increasing rapidly since approximately 2005 and would be projected to surpass hydroelectric in 2016, whereas photovoltaic solar did not begin its dramatic increase until around 2010.

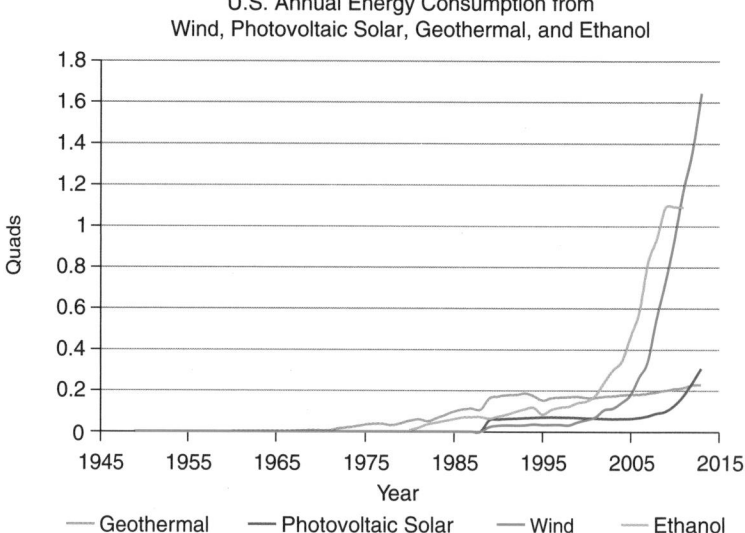

Figure 1.2 U.S. annual energy consumption from wind, photovoltaic solar, geothermal, and fuel ethanol. Geothermal power has been fairly flat since 1990. Wind has been increasing dramatically since 2005, and solar PV began a dramatic increase in 2010. Despite wind's rapid increase since 2005, it still represents less than 2 percent of U.S. power. Similarly, solar PV represents less than 0.4 percent despite a meteoric rise since around 2010.

because the energy used depends on how much time the lightbulb spends lit since the energy consumed is equal to the rate of energy consumption multiplied by the time. Thus, an energy source that delivers 1 watt provides 1 joule per second $\times$ 60 seconds, equaling 60 joules per minute or 3,600 joules per hour, so a 100 W lightbulb uses 360,000 joules an hour.

The total average annual energy use for the world is about 500 exajoules, which is equivalent to an average consumption rate of approximately 2×10^{13} watts. The surface area of the earth is 5×10^{14} square meters. At an average solar flux of about 170 watts per square meter, the sun delivers an average total power of almost 10^{17} watts. Thus, the sun delivers energy approximately 4,000 times faster than we use it. This is not only more than enough for our current needs, but it would be more than enough for future needs even if the world population were to increase by a factor of ten and everyone used energy at the average U.S. rate of 10 kilowatts per capita, rather than the world average of 2.5 kilowatts per capita. Thus, the real issue in renewable energy is not whether the supply is sufficient. The real issue is how we convert enough solar energy into a form that people find useful at a cost that people can afford.

When most people think about solar energy, they think about photovoltaic solar (PV) cells, usually simply called solar cells. Solar cells convert sunlight to electricity, which we can use for almost all of our energy needs. If we were to use solar cells to generate electricity, 100 percent efficient cells would have to cover about 0.02 percent of the entire surface of the earth, including oceans, in order to obtain all our energy directly from the sun. Fortunately, solar cells are not the only way that we can harvest solar power: wind, hydroelectric, and biomass power also convert solar energy into a form that people can use. At the deepest level, renewable energy sources transform energy from sunlight into a form that people can use because sunlight drives the water evaporation the underlies hydroelectric power, as well as the temperature differences that produce wind power and the photosynthesis that allows plants to store energy that can be released by burning plants (e.g., wood) or their by-products (e.g., ethanol). In addition to the benefits of the virtually infinite supply of renewable energy sources, using renewables doesn't increase the net amount of carbon dioxide in the atmosphere. With the exception of systems that burn biomass, renewables do not emit other pollutants, such as smoke, soot, and the chemicals that cause acid rain, as does burning fossil fuels.

Given that renewable energy sources are sustainable, environmentally favorable, and climate friendly, many people suggest that the world should rapidly convert to an economy that depends only on renewable energy even before we actually "run out" of fossil fuels; however, significant economic and social pressures resist such a change. Social pressures are beyond the scope of this book, but throughout the book I will consider some economic issues involved in energy choices.

CAN WE AFFORD TO CHANGE?

Considerations of whether it is economically feasible or even advantageous to build a renewable energy–based power plant rather than a fossil fuel–burning plant require comparisons between very different cost structures. The issue can be illustrated by the simple example of a home owner who is considering putting a solar panel on the roof of her house. If she now pays $100 a month for her electricity, then her annual electricity cost is $1,200. For simplicity, let us assume that for $12,000 she can purchase a solar panel that will provide 100 percent of her average electricity needs. In other words, once she has bought the solar panel, her electricity will be free. Thus, if the electricity price remains constant, she will save $1,200 per year and the solar panel will have paid for itself after ten years. If the solar panel continues to work for the expected twenty years, then she will have $12,000 more than she would have had if she had not bought the solar panels, which corresponds to an annual rate of return of around 3.5 percent even though she would not begin to realize this positive rate of return until after the first ten years. If the residential electricity price doubles during the first five years, however, she will begin profiting sooner. In contrast, if residential electricity price halves during the first five years, buying the solar panel may never be financially beneficial. Thus, whether the solar panel purchase is economically beneficial depends on three factors: (1) the lifetime of the solar panel; (2) the residential price of electricity during the lifetime of the solar panel; and (3) the interest rate paid on capital during the lifetime of the solar panel.

Similarly, for commercial power-generating plants, the initial cost of a fossil fuel–burning power plant is much lower than the cost of a comparable renewable plant; however, in order to generate power, the fossil fuel–burning plant must continue to purchase fuel, whereas the renewable energy plant does not require additional purchases. As a result, whether renewable power is favorable will depend on the price of fossil fuel during the lifetime of the renewable energy plant, which is expected to be twenty to thirty years. Thus, an accurate calculation of the present economic advantage gained by choosing to build a renewable energy power plant rather than a fossil fuel–burning plant requires correctly predicting the price of fossil fuel over a time period equal to that shown in Figure 1.3. The graphs clearly illustrate the enormous difficulty of meeting that predictive challenge given the political and technological changes that can occur during that time period. As discussed previously, the large spike in oil price in 1973 was associated with political events in the Arab world, while the decrease in U.S. natural gas prices after 2005 is largely the result of the supply increases resulting from the technological development of fracking. Neither of these probably could have been predicted ten years before they occurred.

It is important to note that the consumer price of electricity has been much more stable than the fossil fuel prices, which makes consumer solar panel decisions easier. However, someone making a solar panel decision in 1985 would have made much less money than she would have predicted because the price of electricity decreased by almost a factor of two between 1985 and 2005.

Quick calculations based on the 2013 price of energy suggest that a transition to all renewables would increase the price of energy; however, that calculation does not include gains resulting from eliminating inefficiencies associated with fuel burning. Furthermore, the present price of energy may not really accurately capture the full costs associated with energy consumption. Figure 4.2 suggests that simply adding in the current estimated carbon cost would make wind energy the lowest cost source, though wind and natural gas would still be quite close. If additional costs associated with long-term impacts owing to other fossil fuel use consequences, such as ocean oil spills and acid rain, are included, renewables become an even more favorable choice.

The exact long-term physical effects of pollution and climate change are not yet fully known, and the economic and social consequences of

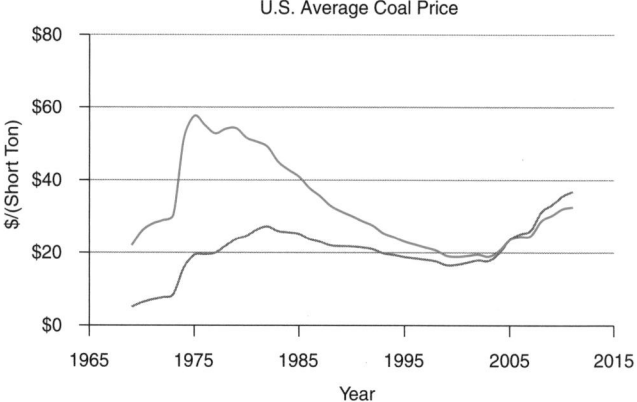

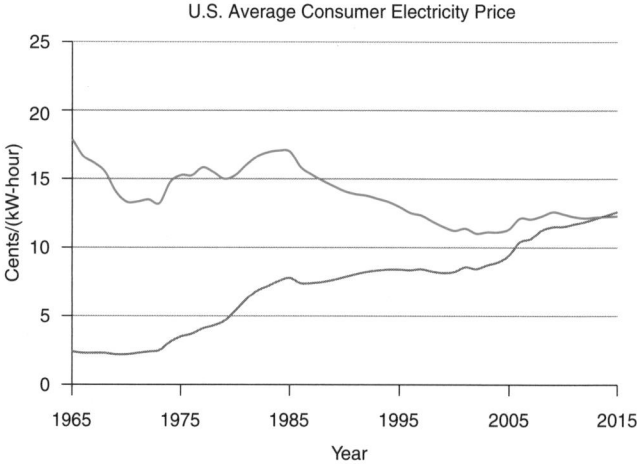

Figure 1.3 Nominal (gray) and real inflation adjusted to 2010 dollars (red) of fossil fuel and consumer electricity prices in the United States, Mcf is an abbreviation for 1000 cubic feet of natural gas. Based on information available in topic 1.6 at www.eia.gov/totalenergy/data/monthly/#summary. The oil prices shown are for imported oil.

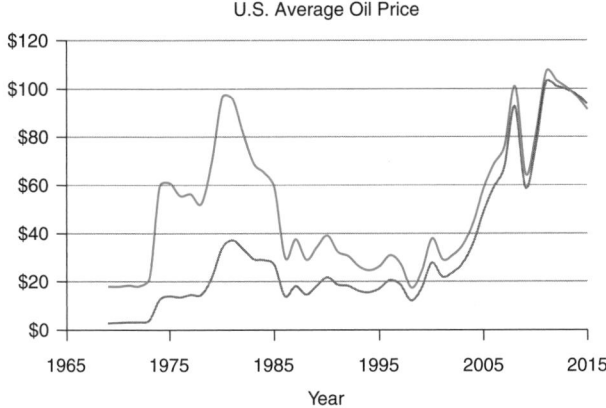

U.S. Average Oil Price

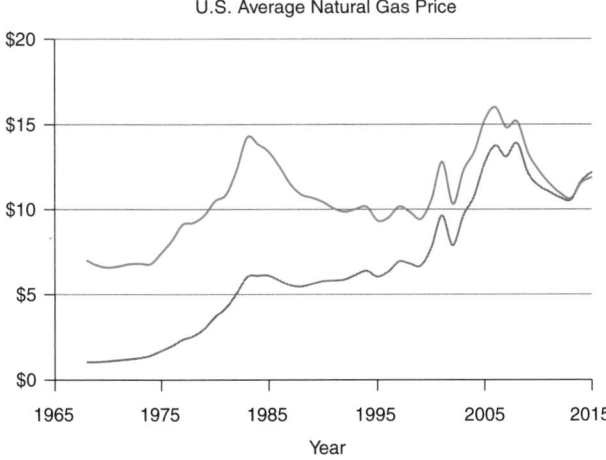

U.S. Average Natural Gas Price

the physical changes are even more uncertain. These costs associated
with long-term impacts highlight again the fundamental issue in converting to renewables. Great economic and social gains may eventually result,
but they will result in a future that is ten to one hundred years away, and
cost projections for such a length of time are extraordinarily unreliable;
however, many people are making renewable energy decisions now.

In sum, a determination of the "correct" price of energy is well beyond
the scope of this book, but I will provide information that will permit

people to compare different energy use strategies, including some consequences associated with energy harvesting and the effects that linger after
the energy is "consumed." I will also discuss whether particular consequences are required by the basic laws of physics or could be altered by
technological developments. I provide a brief example below by considering energy consumption systems in which wood is burned to generate heat.

WOOD BURNING AS AN EXAMPLE OF ENERGY CONSUMPTION

In Africa, approximately 90 percent of the entire continent's population
still uses wood fuel for cooking, and in Sub-Saharan Africa, firewood
and brush supply approximately 52 percent of all energy sources,[4] but
Africa is suffering extensive deforestation, partly as a result of this fuel
wood consumption. Of course, sustainable forestry is quite possible, but
at present much of the wood used for fuel is not obtained from such
sources. People obtain wood by chopping down existing forests because
that wood is "free" or inexpensive, whereas wood obtained from sustainable harvesting is more expensive because its price includes costs associated with planting the trees and maintaining ownership of the land on
which the trees grow. Thus, the people harvesting "free" wood are making a logical immediate choice that benefits them, while ignoring the
future costs to themselves and others that are associated with "free" wood
use today.

Furthermore, though wood smoke has pleasant nostalgic connotations,
it can have adverse effects on human health.[5] Though people might
assume that the negative effects of wood burning are confined to the
developing world, where wood is the major energy source, even in the
developed world wood burning can be the dominant source of particle

---

4. web.mit.edu/africantech/www/articles/Deforestation.htm. Deforestation in
Sub-Saharan Africa Yvonne Agyei. The website states that the article originally appeared
in *African Technology Forum,* Volume 8, Number 1.

5. State of Oregon Publication no. 91-br-023 (revised July 2012): 1–7.

pollution during the winter. A dramatic example is provided by the Australian state of New South Wales, where wood heaters produce up to seven times as much particle pollution as cars.[6] Even in the San Francisco bay area, wood smoke is the leading cause of wintertime particle pollution, with 1.4 million fireplaces and woodstoves contributing about one-third of the region's harmful airborne soot.[7] In such cases, energy consumers may be deciding that the immediate local pleasure resulting from a cheerful wood fire is more important than longer-term deleterious effects associated with the spreading of harmful soot over a large area; however, it is more likely that most people burning wood are simply not aware of the impact of the resulting soot and that some would make different choices if they were aware of the consequences.

As with many areas of energy use, it is not just the energy source that matters. The particular technology used to burn wood also has a profound impact on emissions and energy efficiency. Furthermore, improvements in energy efficiency decrease the amount of harvesting required, which reduces any negative impacts associated with that harvesting. The useful energy obtained from burning wood and the negative consequences associated with wood burning vary tremendously depending on the system that is used. Thus, one cannot simply discuss "wood burning." One must be more specific and consider the system in which the wood is burned.

Open hearths and campfires are romantic, but they are inefficient and generate a lot of air pollution. The graph in Figure 1.4 indicates that open hearth fireplaces emit approximately twenty-eight times more fine particles than an Environmental Protection Agency–certified wood stove and more than 3,000 times more than a gas furnace.[8] These differences

6. See www.environment.nsw.gov.au/woodsmoke/index.htm#contrib.

7. See saratoga.patch.com/articles/air-district-asks-public-not-to-burn-wood-on -christmas-eve-d2da93a0.

8. Coal-burning power plants generate emissions of approximately 2.8 pounds per billion British thermal units, as shown in Figure 10.3. Thus, using an Environmental Protection Agency–certified woodstove to generate heat produces fewer particulates than an electric heater powered by a coal-burning power plant.

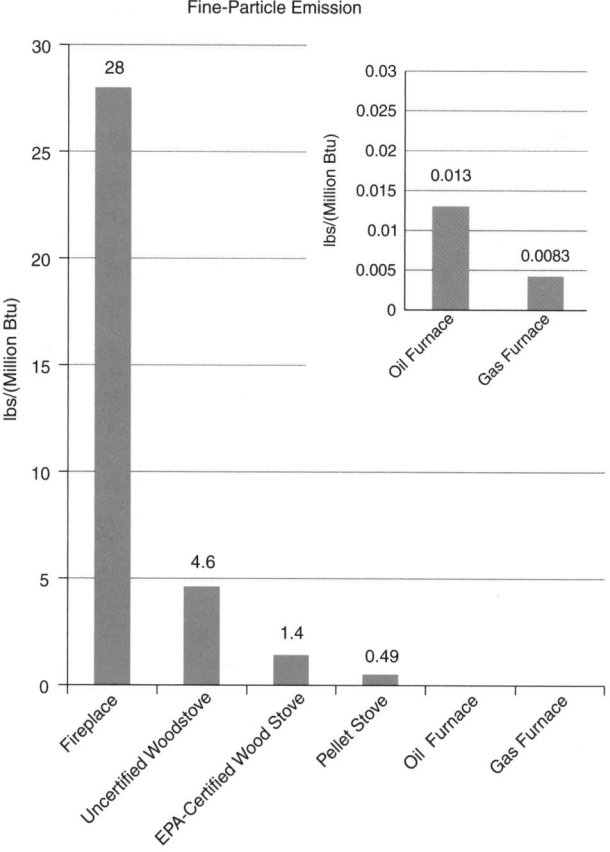

Fine-Particle Emission

Figure 1.4 Emission of fine particles by various systems that burn fuel to generate heat. The numbers above the bars represent the numerical values. The inset is the graph for the oil and gas furnaces where the y-axis is expanded 1,000 times. Thus, the fireplace generates more than 3,000 times the fine particle emission of a gas furnace providing the same amount of heat. Available at www.epa.gov/burnwise/energyefficiency.html, the data is from the bar graph.

in particle emission provide a very dramatic example of a case in which switching to renewables can have negative environmental consequences. Whether the overall environmental impact of replacing natural gas with wood is positive or negative would involve an extremely complex calculation that would have to include various factors, including the long-term effects associated with carbon emission which are quite uncertain at present.

Not only do particle emissions vary strongly depending on the wood burning technology used, but efficiencies vary strongly as well. At worst, even though a person standing directly in front of an open burning fireplace feels warm, such fireplaces draw warm air from the room and send it up the chimney, resulting in a net heat loss to the house, though a person standing directly in front of the fireplace still feels warm. At best, open fireplaces are only about 10 percent energy efficient. Wood burning stoves with enclosed chambers avoid forcing warm air out of the house, resulting in increases in energy efficiency. As a result of such technological improvements, new models claim energy efficiencies up to 60 to 80 percent, which seems spectacular; however, these numbers still don't best gas furnaces that offer efficiencies of 75 to 98 percent based on their Annual Fuel Utilization Efficiency (AFUE) ratings. Thus, not only does nonrenewable natural gas produce fewer emissions than wood burning, it is also more energy efficient than wood burning.

Some care must be taken in interpreting energy efficiency numbers. The 60 to 80 percent numbers for modern wood burning stoves assume that the system is operating at peak heat generation. If less heat is being generated, the efficiency is lower. The tradeoff between efficiency and peak capacity that occurs in wood stoves is common to many energy systems. In the case of wood burning, there are two major options: (1) make the size large enough to meet the peak heat requirement, which implies that most of the time it will be operating well below its theoretical efficiency; or (2) size the system to equal the average load rather than the peak load, which means that during times of peak demand the system will not be able to provide enough heat. In the latter case, the overall efficiency of the system will improve, but inadequate heat will be provided on very cold days. That problem could be offset by having two or three wood stoves, where the number of stoves used is determined by the total amount of heat required, but that solution, while more energy efficient than one stove, is economically inefficient and takes up a lot of space. The same issues associated with the tradeoff between capacity and efficiency that exist for wood burning energy also arise for most other power systems in use or development.

For example, electric power companies must provide peak power on maximum air conditioning demand days, even though the average demand is much smaller. As a consequence, power companies must either suffer significant capital inefficiencies by maintaining generators that are rarely used or risk brownouts or blackouts during periods of unusually high demand.

Capacity versus efficiency trade-offs such as these highlight the benefits that could be obtained if we possessed efficient, compact, inexpensive energy storage. Such energy storage would allow us to design systems that operate near peak efficiency all of the time. When the heat or electricity generated exceeds demand, the excess energy could be stored and used during periods where the demand exceeds the generating capacity. At present, no such system is available; however, it may be the area where a revolution in science or technology would have the largest impact on the economics of energy use regardless of source, while also making the transition to renewable energy sources far more feasible than it is at present given that the energy generated by both wind power and solar power varies enormously with time.

CHOICES

Energy consumption choices are not made by physicists and engineers; they are made through complex interactions among psychological, social, and economic factors. All of us make many separate individual decisions about energy consumption every day, and as voters we also have an effect on government policies concerning energy use. Any given choice seems insignificant, but in aggregate these choices determine how energy is used. At present, individual consumers as well as national and international policy makers are rapidly altering energy use. For example, as a result of the economic crash in 2008, the government of Ireland instituted a carbon emission tax to raise revenue in an effort to balance its budget. According to a *New York Times* article published on December 27, 2012,[9] largely as a result of decisions made by individual consumers in response to the tax, carbon emissions in Ireland have dropped 15 percent during the last four years, and the tax provided around 25 percent of the additional revenue needed to balance the budget, while subsidies were still provided to offset the impact of the tax on low-income families. Furthermore, in response to consumer decisions, corporations such as Mercedes Benz reduced the carbon emission of products that they market in Ireland in order to make them more attractive to Irish consumers. In the United States, it is estimated that a tax similar to that instituted by the Irish government could reduce the U.S. budget deficit by 50 percent over a ten-year period.

The United States does not at present tax carbon; however, U.S. companies are already doing internal planning using priced carbon under the expectation that carbon will eventually be taxed.[10] According to the *Economist,* ExxonMobil prices carbon dioxide at $60 a ton, BP and Shell use $40 a ton, and Microsoft uses $6–$7 a ton. In the European Union, the market price is $6.70 a ton, and in California it is $11.50 a ton. The United States estimates that the social cost of carbon dioxide emission is $37 a ton.

Even in the absence of a carbon tax, the United States is in the midst of an energy revolution as coal is replaced by natural gas because of the decrease in natural gas prices stemming from the increase in gas supply

9. See www.nytimes.com/2012/12/28/science/earth/in-ireland-carbon-taxes-pay -off.html?partner=rss&emc=rss&src=ig&_r=0m.

10. The *Economist,* December 14, 2013, www.economist.com/news/business /21591601-some-firms-are-preparing-carbon-price-would-make-big-difference -carbon-copy. Some firms are preparing for a carbon price that would make a big difference.

resulting from fracking. This market-driven decision is resulting in a significant reduction in power plant pollution, including carbon dioxide generation, which has been reduced by 20 percent in the period between 2008 and 2012, achieving levels last seen in 1992.[11] Similarly, shale oil may drastically lower the U.S. dependence on oil imports without direct government intervention in the form of gasoline taxes or import tariffs.

Though much discussion of hidden costs is directed at fossil fuels, renewables have substantial hidden costs as well. The most obvious hidden cost is government subsidies. The form of these subsidies varies. For example, in the United States, there are tax deductions for purchasing solar equipment, whereas in Germany the capital expense is not subsidized; however, subsidies are paid for power fed from solar panels into the power grid. Wind power and ethanol production also receive U.S. government subsidies. Between 2005 and 2009, the ethanol subsidy was approximately $6 billion per year.[12]

Government regulation can also introduce hidden costs without subsidies. One example is provided by the introduction of ethanol into gasoline in the United States, which was required by U.S. government regulation.[13] One might assume that the government would encourage this regulation to be met in the most cost-effective manner; however, this is not the case. Ethanol made from sugar cane in Brazil and deliv-

11. See www.eia.gov/todayinenergy/detail.cfm?id=7350.

12. See usnews.nbcnews.com/_news/2011/12/29/9804028-6-billion-a-year-ethan ol-subsidy-dies-but-wait-theres-more?lite.

13. The timeline for phasing in enormous volumes of renewable fuel was laid out in the Energy Independence and Security Act of 2007. It requires that a certain total amount of ethanol be included in U.S. gasoline. In 2013, 16.55 billion gallons were required to be added, while 18.15 billion gallons would be required in 2014, which would make that ethanol content of gasoline exceed 10 percent, approaching the percentage where vehicle damage due to ethanol combustion becomes significant. Thus, in 2014, the Environmental Protection Agency (EPA) is suggesting that the required amount of ethanol be reduced to between 15 and 15.5 billion gallons. See money.cnn.com/2013/11/15/news/economy/epa-ethanol/.

ered to the United States has been significantly cheaper than ethanol made in the United States from corn. To discourage imports, the U.S. government imposed a large tariff on imported ethanol that made U.S.-produced corn ethanol less expensive than ethanol imported from Brazil. As a result, U.S. consumers paid more for motor fuel containing ethanol than they would have paid had there been no tariff on Brazilian ethanol.

In the case of ethanol, there are additional hidden costs that are even more subtle. The *Economist* stated that "ethanol is the dominant reason for this year's increase in grain prices."[14] It accounts for the rise in the price of maize because the federal government has in practice waded into the market to mop up about one-third of America's corn harvest. A recent study indicates that ethanol production has made corn prices 30 percent higher than they would have been in the absence of ethanol production. More dramatically, on March 25, 2013, Babbage's blog for the *Economist* argued that the retail price of gasoline in the United States is now being driven up by federal requirements on the amount of ethanol that must be included in gasoline.[15]

Adding ethanol to gasoline also adds an additional even more subtle price increase: the energy per gallon for ethanol is only two-thirds of the energy per gallon of gasoline. Thus, one gallon of E85 gasoline has only about 70 percent of the energy contained in one gallon of pure gasoline. This represents a secret energy price increase of 40 percent of which most consumers are unaware. Increased automobile maintenance costs associated with ethanol-related damage to vehicles not specifically designed to consume ethanol can add yet another hidden cost for some

14. "The end of cheap food: Rising food prices are a threat to many; they also present the world with an enormous opportunity," *Economist,* December 6, 2007, www.economist.com/node/10252015.

15. "Difference Engine: End the ethanol tax," *Economist,* March 25, 2013, www.economist.com/blogs/babbage/2013/03/biofuels.

vehicles. The problems associated with 15 percent ethanol are significantly higher than those associated with 10 percent ethanol.[16]

Cost increases are not the only negative impact associated with adding ethanol to gasoline. The transition from gasoline to gasoline-ethanol mixes is an example of how converting to renewables increases the emission of some pollutants: ethanol burning releases approximately twice the ozone released by burning the equivalent amount of gasoline.[17]

There were several motivations for the passage of the Energy Independence and Security Act of 2007, which requires the addition of ethanol, including reducing dependence on oil imports, reducing net carbon emission, and providing an economic boost to regions in the United States where corn is grown. According to a November 15, 2013, article in the *New York Times,* in the years since the law was passed, imports of oil have declined sharply, domestic production has risen, and consumption has also declined.[18] The 2007 law set ethanol quotas that would have meant a modest percentage increase in the amount of ethanol used in the mix had consumption continued to increase at the rates predicted in 2007, before the financial crash. Instead, the percentage of ethanol in gasoline has increased significantly and is now approaching 10 percent, which is the maximum ethanol allowable in most gasoline because vehicle damage becomes increasingly significant if the fraction exceeds this threshold, unless the vehicle is specifically designed to accept mixed fuel. Meanwhile, increases in domestic oil production are so large that they will reduce oil imports even if no ethanol were added to gasoline.

16. See www.mossmotors.com/SiteGraphics/Pages/ethanol.html; forums.aaca.org/f133/important-ethanol-message-289618.html; engines.honda.com/parts-and-support/fuel-recommendations#e85.

17. See news.stanford.edu/news/2009/december14/ozone-ethanol-health-121409.html.

18. For First Time, E.P.A. Proposes Reducing Ethanol Requirement for Gas Mix, by Matthew L. Wald, Published: November 15, 2013. www.nytimes.com/2013/11/16/us/for-first-time-epa-proposes-reducing-ethanol-requirement-for-gas-mix.html?module=Search&mabReward=relbias%3Ar.

As a result of all of the negative consequences associated with the government-required insertion of ethanol into gasoline, in 2013 a bipartisan group of ten U.S. senators introduced a bill to eliminate the mandate for corn-based ethanol. The senators argue "that current law raises the cost of food and animal feed and damages the environment."[19]

The fuel ethanol transition in the United States also illustrates that decisions made about energy have an influence that extends far beyond direct consequences associated with energy use and extends well beyond the borders of the United States. In 2011, 38 percent of the U.S. corn harvest was diverted to ethanol production, resulting in a 3.3 percent reduction in global grain production.[20] In "The Effect of the U.S. Ethanol Mandate on Corn Prices," Carter, Rausser, and Smith state that those price increases "have had particularly devastating consequences for consumers in less-developed countries, where a relatively large percentage of income is spent on food, and where grains, rather than processed foods, constitute the major portion of the diet."[21] Thus, the increase in U.S. ethanol production resulting from a government mandate is an example of an energy decision made in one country that has had a large impact on other countries, because the increase in U.S. ethanol production increased world grain prices.

In sum, though governmental decisions about energy use, such ethanol mandates, carbon taxes, and emission standards tend to be made at the state or national level, the consequences of these decisions can strongly affect people in other countries, often in ways not foreseen by the original decision makers. The fact that one cannot possibly anticipate all the consequences of a choice does not imply that one should make no choices; however, it does suggest that some care should be taken to consider not

19. See www.reuters.com/article/2013/12/12/us-usa-ethanol-corn-idU.S.BRE9B B0WR20131212.

20. "The Effect of the U.S. Ethanol Mandate on Corn Prices," agecon.ucdavis.edu /people/faculty/aaron-smith/docs/Carter_Rausser_Smith_Ethanol_Paper_Sept_2013.pdf.

21. Ibid., 2.

only obvious short-term consequences but also subtle long-term consequences that may eventually be the dominant outcomes of our decisions.

In most cases, energy consumption choices will not be driven by physics or engineering, but the choices must still accord with the laws of physics. In particular, energy conservation is a basic law of physics, which holds that energy can never be created or destroyed. There will never be more energy than there is now, but similarly there will never be less. Though energy can never be created or destroyed, it can be repeatedly transformed from one type into another; however, each transformation always involves some loss. The amount of loss depends on the details of the process. Both fundamental physical limits and engineering flaws contribute to losses, but only the latter can be changed. Most economic and political discussions of energy focus on the transformations between energy sources—coal, oil, gas, solar, wind, nuclear, and biomass power— and what people deem end uses, such as transportation, climate control, and manufacturing. Of course, the energy existed in some form before it was incorporated in coal, and it exists in some form after we are finished using it for our purposes.

In order to make reasonable choices, various aspects of energy use should be considered to determine which particular system, for example, fireplaces versus woodstoves, results in the most favorable overall outcome, including not only what we immediately experience, such as the beautiful flame of an open fire, but also what changes occur before and after we enjoy that beautiful flame. Thus, for each particular energy consumption system, it is important to consider the following: (1) the effects associated with energy harvesting, for example, deforestation; (2) postconsumption consequences, for example, smoke pollution; and (3) energy efficiency versus capital efficiency tradeoffs, for example, sizing the system for peak versus average consumption.

The next few chapters of this book will discuss the basic science underlying various energy exploitation systems, including a detailed discussion of effects associated with harvesting and use, as well as trade-offs between capacity and efficiency. I will begin by considering electricity and then move on to consider energy conservation, energy storage, and consequences of consuming energy. The consequences of particularly important energy decisions are really about creating the best possible overall outcomes, where all of the implications of energy choices are considered.

This book includes three appendixes covering the Carnot limit, electricity generated by heat, and some recommendations for the future. Several additional appendixes are available online at http://thedata.harvard .edu/dvn/dv/HUP. Online Appendix 1 provides a detailed qualitative discussion of the analogies between currency and energy in order to highlight qualitative features of the scientific rules governing energy use. Online Appendixes 2 and 3 discuss scientific measurements and the units used to make those measurements.

2

ELECTRIC POWER FOR
A RENEWABLE FUTURE

Unlike wood and oil burning, which have been traditional methods of providing energy for thousands or hundreds of thousands of years, electricity use on a commercial scale is a modern phenomenon that has become important only in the last hundred years. Part of the reason that electricity was not used earlier is that natural "sources" of electricity such as lightning, electric eels, and shuffling across carpets are not easily harnessed to meet human energy needs, though lightning is sufficiently awe inspiring to have made substantial contributions to art and religion for millennia. Several myths argue that lightning provided man with the divine gift of fire; however, this chapter will show that in the developed world, fire provides us with the gift of lightning, or at least the electricity that is delivered to consumers over the massive, complex, and expensive power infrastructure that has been created over the last 150 years.

Generating electricity in a form that people can use required a significant body of fundamental scientific knowledge that was not available until the mid-1850s. The development of a national electrical infrastructure is a triumph of basic science and engineering that also required enormous capital investments. Traditionally, much of the power generation had been located within tens of miles of the consumers. Locating power plants near consumers is sensible because transmitting electrical power along wires always entails an energy loss—the amount of power generated by fossil fuel burning power does not depend on the physical location of the plant. The energy loss increases with the distance transmitted, so locating power plants close to consumers minimizes energy losses resulting from transmission. Thus, many U.S. cities have electrical power plants within the city limits that provide electricity for the city. As a result, power transmission occurs often over miles or tens of miles. In contrast, the performance of renewable power plants does depend strongly on the location of the plant. For example, the amount of wind power generated by a wind turbine depends on how much wind is present where the turbine is located; therefore, it is advantageous to locate wind turbines in windy locations that are often distant from the population centers where most electricity is consumed. Thus, the transition to renewable energy will create the need for transmission distances extended hundreds of miles, but it will also provide energy within less than a hundred yards of where it is consumed; however, in a renewable world the need for transmission on the scale of tens of miles may well disappear.

The increased transmission distances arise because the power generated by some renewable sources depends strongly on location. For example, wind farms perform better in locations that are windy. In addition, many people find large wind farms displeasing on various grounds, including aesthetics. As a consequence, in the developed world, an expensive and complex new long-distance power delivery infrastructure is being created to link remotely located renewable energy sources, such as wind farms and hydroelectric dams, to urban areas with large populations. Power transmission lines in these systems typically extend hundreds of miles, so transmission losses are a much greater concern for renewable power plants located far from customers than they were for local power plants separated from customers by about ten miles. In Chapter 7, I will discuss some of the technical challenges associated with such long-distance transfers and why such long-distance transfers are so different from the short-distance transfers that are at present the norm.

In contrast to the long- and increasingly longer-distance energy transfers required in the developed world, in the developing world renewable energy in the form of solar panels offers electric power generation located within a few meters or even centimeters of the device that consumes the electricity. Such systems require very little capital and provide electricity in the complete absence of infrastructure. Even in the United States, garden lights consisting of one single unit with solar panels on top and light-emitting diodes just below them have become popular as flexible, reconfigurable lighting sources that do not require any connection to the U.S. power grid. More seriously, solar panels on the roofs of U.S. houses can provide an average power that exceeds the average electricity demand of the household, as will be discussed in detail in Chapter 5.

At present, most people in the developed world use electricity to provide cooling, lighting, and power for electrical devices, as illustrated in Figure 2.1. This graph shows that cooking, refrigeration, and water heating account for 29 percent of power consumption, while air conditioning accounts for 21 percent and lighting for 14 percent. The evolution of air conditioner efficiency is one of the reasons that energy use per capita in the United States has not increased in decades despite increases in the size of living spaces. As a result of technological improvements, current air conditioning power requirements are only around 50 percent of the power requirements for air conditioners built in the 1970s. In other words, the new systems provide exactly the same quality of life using one-half the energy. The theoretical limit for such systems is approximately ten times more than the efficiency of current systems, and existing commercial systems are available with twice the current standard efficiency; consequently, this is also an area where technological improvements will continue as a result of both commercial pressure and government regulation, including new minimum efficiency standards that will come into effect in 2015.

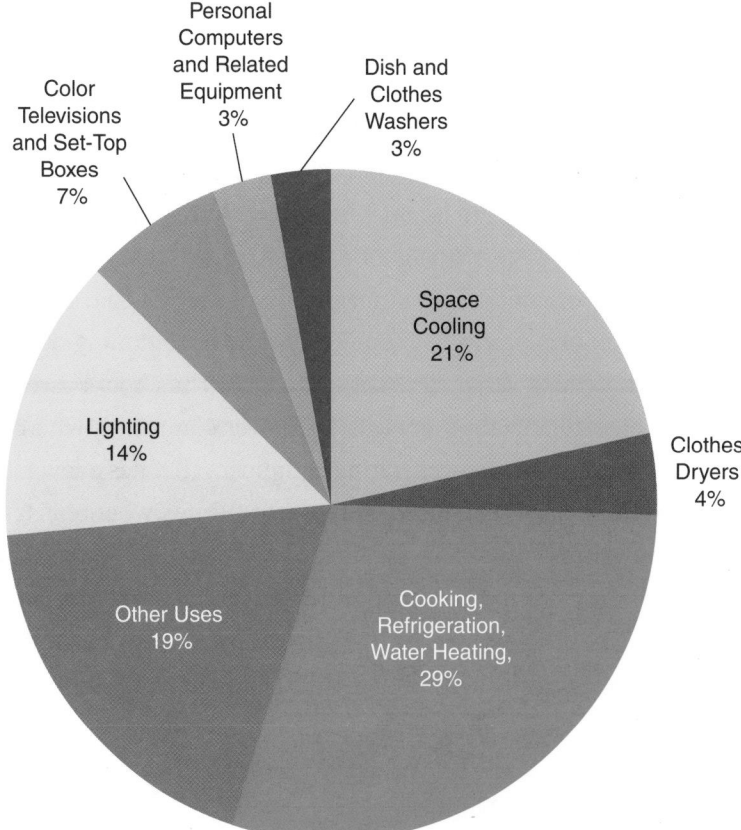

U.S. Residential Electricity Use 2011

Personal Computers and Related Equipment 3%

Color Televisions and Set-Top Boxes 7%

Dish and Clothes Washers 3%

Space Cooling 21%

Lighting 14%

Clothes Dryers 4%

Other Uses 19%

Cooking, Refrigeration, Water Heating, 29%

Figure 2.1 U.S. residential electricity use in 2011, from U.S. Energy Information Administration data. Available at www.eia.gov/forecasts/aeo/excel/aeotab_4.xlsx. In general, table A4 at www.eia.gov/forecasts/aeo/ provides current versions of this information.

The large and fluctuating nature of the demand for air conditioning is an example of a situation where the fluctuating nature of renewable energy can be advantageous: days with high air conditioning demand are also days where lots of solar power is provided. Thus, solar panels provide the most power when the most power is needed. In Chapter 5, I discuss solar power in detail. To provide a brief example here, at noon in Arizona existing commercial solar panels can provide around 200 watts of electricity for each square meter of solar panel.[1] A suburban roof with a hundred square meters of solar panels would produce 20,000 watts of power at noon, which is enough to provide about 2.5 times the power required to air-condition a 3,000 square-foot house.[2] Thus, the panels could not only power the house on which they were located, they could provide complete power for a neighbor as well, even if that neighbor was also running the air conditioner at maximum power. Of course, at midnight the solar panels produce no power, but the conditioning demands are smaller, as are most other electricity demands. Thus, roof-based solar panels may offer financial advantages for power companies who could eliminate existing fossil fuel–generating equipment that has a low return on investment because it is used only during times when electricity demand is very high.

Figure 2.1 indicates not only opportunities for conservation but also the limitations of different energy reduction schemes. For example, the conversion from incandescent lighting to LED lighting is occurring fairly rapidly as a result of a number of different factors. Many people are arguing that the switch will result in a large decrease in energy consumption;

1. At noon, the delivered solar power is approximately 1,000 watts per square meter. Commercial solar panels are around 20 percent efficient, so the electricity provided would be 200 W/m^2 × 100 m^2 = 20,000 W or 20 kW. Residential central air conditioners consume from 3 to 20 kilowatts of power. Air conditioners with an Energy Efficiency Ratio (EER) rating of 8 require approximately one-eighth of a watt per Btu to operate. On their website at energy.gov/energysaver/articles/room-air-conditioners, the United States Department of Energy suggests that around 20 British thermal units per square foot is required. Thus, a 3,000-square-foot house would require 60,000 British thermal units, which would consume 7.5 kilowatts of power. The 20 kilowatts from the solar panels is more than 2.5 times that.

2. Of course, the energy required by different houses depends on the insulation in the house, the energy efficiency of the air conditioner, and the temperature at which the thermostat is set. For example, new air conditioning units have EER of 10, so 20 kilowatts would provide 200,000 British thermal units, which would power the air conditioners of more than three houses.

however, the graph shows that even eliminating lighting would reduce residential electricity consumption by only 14 percent. Additionally, the graph also shows that technological changes do not always result in reductions in energy, as evidenced by the fact that televisions and cable boxes now consume as much energy as dishwashers, clothes washers, and clothes dryers combined.

Though the majority of the electricity sold in the United States is sold to residential customers, commercial use of electricity is almost as high, as one can see by inspecting Figure 2.2. Figure 2.2 indicates that industrial use has remained more or less constant over the last ten years. Though the amount of electricity used by industry has remained constant, overall use of electricity has increased; therefore, the fraction consumed by industry has decreased, though the actual amount of electricity consumed has remained constant. In contrast, commercial use has increased by slightly more than 20 percent, and residential use has increased by slightly less than 20 percent. As a result, the fraction consumed by both of those sectors has increased during the last ten years. The fraction of electricity consumed by transportation is too low to be shown on the graph; however, in the near future, as more electrical automobiles come into service, the contribution of electricity to transport is likely to increase. It is important to note that the electricity generated and used in hybrid cars would not appear on this graph because that electricity was never "sold" as retail electricity. Similarly, electricity generated and used by domestic solar panels does not appear. In the future, these personal sources of electricity may play a significant role in altering the relationship between consumers and commercial energy providers.

Calculations of the feasibility of replacing fossil fuel–generated electricity with electricity generated from renewable sources depends on the cost of the former. Figure 2.3 and Table 4 show that the cost of residential electricity is not the same in all U.S. locations. The large variation in the price of electricity by state means that comparisons of the cost of

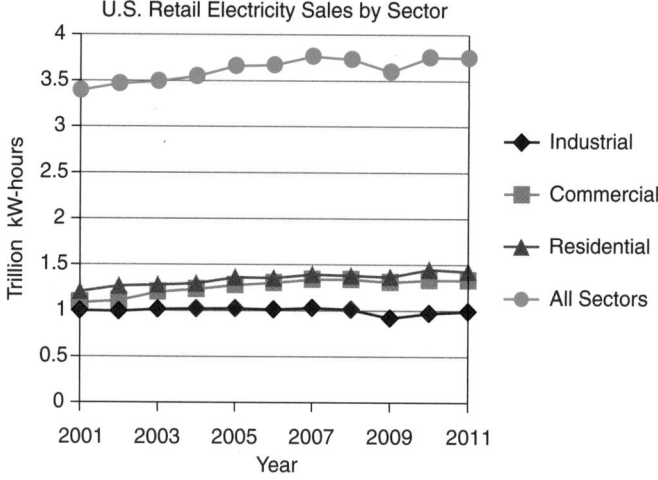

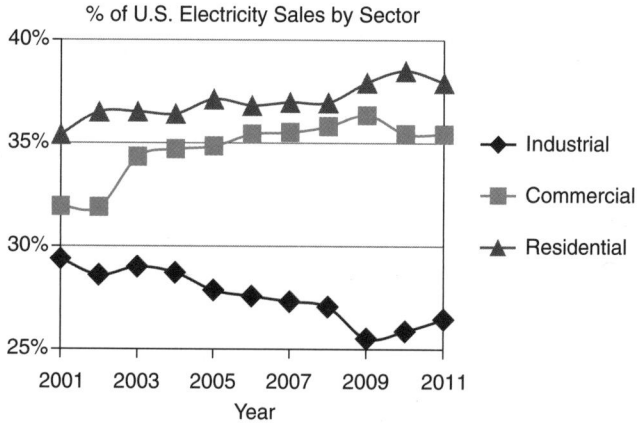

Figure 2.2 U.S. retail electricity sales by sector for the years 2001–2011, from U.S. Energy Information Administration data available as topic 7.6 at www.eia .gov/totalenergy/data/monthly/#electricity.

renewables and the national average price of electricity may miss significant differences between states. Such differences are, for example, extremely important for someone trying to decide whether to put a solar panel on the roof of his house. In this book, I will consider the national averages, but the information in Table 4 can be used to provide comparisons for

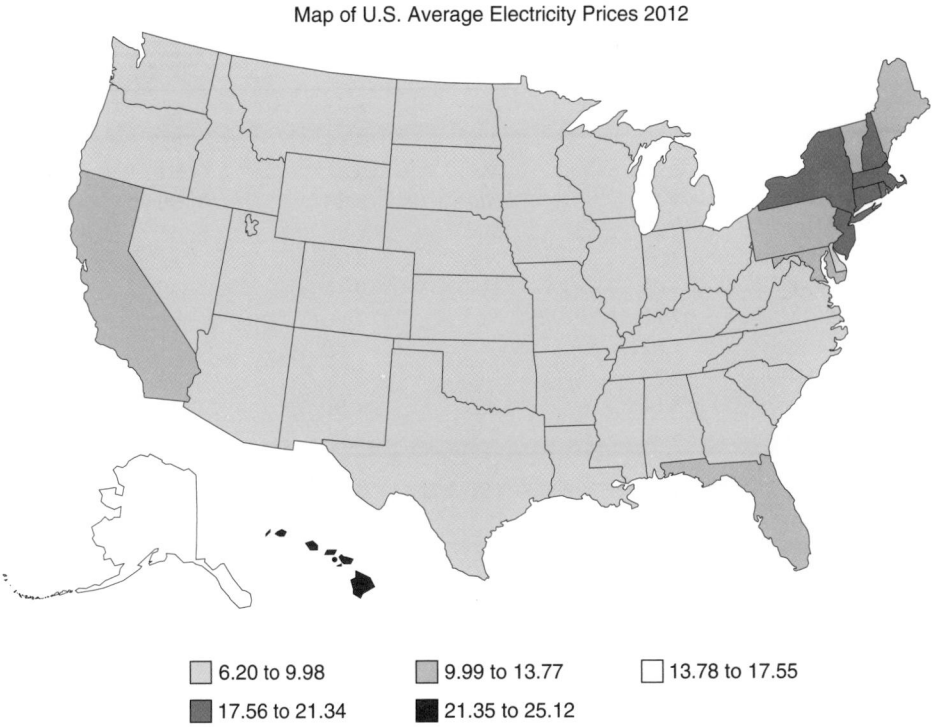

Figure 2.3 U.S. electricity price by state. National average is 9.8 cents. Prices in Hawaii are more than twice the price in the next-highest states. From U.S. Energy Information Administration 2012 data. Current and past data is available at www.eia.gov/electricity /monthly/epm_table_grapher.cfm?t=epmt_5_06_a.

particular states. Arizona's average use is approximately equal to the national average, so states appearing before Arizona consume less than the national average, while states appearing after Arizona consume more.

As mentioned above, historical use is likely to differ strongly from future use. The quick calculation in Chapter 1 showed that the solar energy delivered to the earth is approximately 4,000 times larger than our current energy needs, which is important because all of our renewable energy originates in sunlight. Given that electric cars are commercially available,

Table 4.   U.S. electricity prices, by state, ordered from lowest to highest.

State	Price	State	Price	State	Price
LA	6.9	MO	8.53	WI	10.28
ID	6.92	TX	8.54	FL	10.44
WA	6.95	MS	8.6	MI	10.98
WY	7.19	NM	8.83	DE	11.06
KY	7.26	MN	8.86	MD	11.28
OK	7.54	NV	8.95	ME	11.81
AR	7.62	VA	9.07	DC	11.85
IA	7.71	SC	9.1	RI	12.74
ND	7.83	OH	9.12	CA	13.53
UT	7.84	NC	9.15	NJ	13.67
WV	8.14	AL	9.18	MA	13.79
OR	8.21	TN	9.27	NH	14.19
MT	8.25	KS	9.33	VT	14.22
IN	8.29	GA	9.37	NY	15.15
NE	8.37	CO	9.39	CT	15.55
IL	8.4	AZ	9.81	AK	16.33
SD	8.49	PA	9.91	HI	34.04

Source: Energy Information Administration 2012 data. The EIA provides monthly data at http://www.eia.gov/electricity/monthly/epm_table_grapher.cfm?t=epmt_5_6_a, which offers both current and historical data.

almost all of the world's energy needs could be met using electricity.[3] If that electricity were generated by inexpensive, local, renewable energy sources, then many of the troubling issues surrounding energy use would be resolved. For this reason, the European Union has promulgated policies designed to encourage the transition to a system in which electrical energy use dominates, while also pushing policies that boost the fraction of electrical energy provided by renewable sources. As a result, Germany and Scandanavia have greatly increased renewable energy–based electricity generation. At present, Denmark generates around 20 percent of its electrical energy from wind power, and by 2020 it intends to raise that percentage to 50 percent. More dramatically, on November 3, 2013, the electricity produced by wind in Denmark exceeded 100 percent of the

3. Biofuels could provide the energy for applications that require combustion, such as jet engines and some industrial processes.

Table 5.   Quantities characteristic of U.S. electricity energy use.

U.S. annual electrical energy end use	3,800,000 MWh/year
U.S. annual electrical energy end use, in joules	1.5×10^{19} J
Average U.S. electricity end use, in watts	4.4×10^{11} W
Population of the United States	320,000,000
Average U.S. electricity end use, in watts/capita	1,500 W/capita
Average U.S. electricity end use, in kW/capita	1.5 kW/capita
Average U.S. electrical power use per residence	1.3 kW/residence
Land area of the United States	9.2×10^{12} m^2
Average U.S. electrical power per square meter	0.044 W/m^2

demand for electricity.[4] Furthermore, on one bright summer weekend day in 2012 the majority of the electricity generated in Germany was provided by solar power.[5] The 22 gigawatts generated by solar power that day is equivalent to the output of approximately twenty nuclear power plants. Of course, on a dark winter weekday, instantaneous solar power provides a much smaller fraction of the energy consumed, and solar power never provides energy during the night. Together, these two examples demonstrate that either wind or solar power can provide a large fraction of the electricity consumed in a major industrial country. In what follows, I will consider the present electricity generation system in the United States. I will then show that renewables could not only easily meet 100 percent of the U.S. demand for electricity, they could meet 100 percent of the total U.S. energy demand if that power was provided in the form of electricity.

Table 5 summarizes the current electricity use in the United States, and Table 6 provides similar information for total energy consumption in order to highlight the challenges of making a transition from our current system to a system where almost all energy is provided by electricity created by renewables sources.[6]

4. See energytransition.de/2013/11/denmark-surpasses-100-percent-wind-power/.
5. See www.reuters.com/article/2012/05/26/us-climate-germany-solar-idU.S.BR E84P0FI20120526.
6. In order to aid interpretation of the information in the tables, online Appendixes 2 and 3 at http://thedata.harvard.edu/dvn/dv/HUP provide a more detailed discussion of SI units, including joules and watts.

Table 6.   Quantities characteristic of total U.S. energy use to allow comparison with current electricity use. The total energy consumption is almost seven times the electrical energy consumption.

Total U.S. annual energy use in all forms	10^{20} J
Average total U.S. power use in all forms	3×10^{12} W
Average U.S. power use in all forms per capita	10,000 W
Average U.S. power use in all forms per square meter	0.34 W/m^2

Of particular use are the expressions in terms of watts per square meter because renewable can be characterized by the number of watts per square meter that can be extracted. The renewable energy that can be harvested within the national boundaries of a country is then the product of the watts per square meter and the surface area of the country. Countries with larger surface areas can harvest more renewable energy, but what matters is whether that total energy is smaller or greater than the total energy consumed by that country. Small countries may provide plenty of energy for their citizens if the country is sparsely populated and/or the energy consumption per capita is small. For countries with high population densities and energy consumptions per capita, a complete transition to renewable energy is not possible without energy imports. For countries where conversion is possible, one can use the energy density to calculate the fraction of land area that must be devoted to renewables to harvest the desired amount of energy.

At first it seems obvious that one cannot devote more than 100 percent of land area to renewable power, but the situation is slightly more subtle. Solar power and wind power can both be harvested from the same land area. Such "double use" of the same land not only increases the average power density, it also decreases the fluctuations in power generation associated with each separate system. As a result, some systems that combine solar and wind harvesting are now commercially available. In addition, since water that falls onto land used for wind and solar power may flow into streams and rivers captured by hydroelectric dams, the same land can provide three types of renewable energy. Thus, the sum of the land area devoted to wind, solar, and hydroelectric power could be more than 100 percent of the total land area since the same land area could contribute to more than one type of renewable energy.[7] Of course, renewable power is not the only important use of land. Agriculture and housing also require land. It is important to note that in the case of hydroelectric power or wind power, most of the land from which the energy is harvested remains available for agriculture or any other purpose. In contrast, present solar power harvesting systems remove all of the sunlight that falls on the area covered by the solar harvesters, producing complete darkness below. Thus, land used for solar power cannot be used for agriculture or any other use that requires sunlight.[8] Solar power is not the only renewable energy source that reduces the amount of land that can be used to raise crops. A tract of land that is being used to grow a biofuel crop cannot be used to raise a food crop at the same time unless the biofuel comes only from agricultural waste. As a result, people are pursuing basic research to try to find efficient methods of extracting biofuel from agricultural waste. The present competition between food crops and fuel crops is not merely a philosophical issue since the conversion of U.S. farmland to cultivation of corn for ethanol production increased the global price of food grains, as discussed in some detail at the end of Chapter 1.

OVERVIEW OF ELECTRICAL POWER GENERATION

Even without a complete conversion to an electrical energy economy, electricity plays an extremely important and almost constant role in the lives of most people in the developed world. When we say that electricity flows along a wire, we mean that there is literally a net flow of negatively

7. The land actually occupied by the water behind hydroelectric dams is not useable for conventional agriculture or housing.

8. However, building roofs already block sunlight, so placing solar panels on roofs does not reduce the amount of land available for other purposes.

charged electrons moving along the wire. The flow of electrons is referred to as a current, in analogy with the flow of water. In SI units, electrical current is measured in amperes, which is usually shortened to amps.

Just as water flows downhill in a gravitational potential, electrons fall downhill in an electrical potential. In SI units, the electrical potential is measured in volts. Thus, a 9-volt battery provides more potential difference than a 1.5-volt battery; if you connect a wire between the terminals of a 9-volt battery, the current flow will be larger than if you connect the same wire across a 1.5-volt battery.

We are rarely conscious of the flow of electricity that supports us; however, when the battery dies or the power stops coming out of the wall, we notice. In developing countries, brownouts and rolling blackouts are normal features of everyday life, but in most developed countries, an electricity outage lasting more than a few hours is a memorable event. The Northeast blackout of 2003 was the second largest blackout on record and affected areas in the northeastern United States and Canada as far west as Ontario. It occurred just before 4:10 P.M. on August 14 and lasted for seven to sixteen hours, depending on the region. It left more than 55 million people without electrical power and shut down a range of basic infrastructure services, including transportation and communication services in addition to air conditioning and lighting. Similarly, at 5:27 P.M. on November 9, 1965, the Northeast and much of Canada lost power, which was not restored until the following morning. Though modern life underwent a serious shock in both cases, neither event spawned significant unrest or produced long-term aftereffects (urban legend regarding a spike in the birthrate nine months after the blackout was false).[9] The 1965 outage was reflected in one of Doris Day's last movie titles, "Where Were You When the Lights Went Out?" Though

9. Internet fact-check and reference source Snopes.com indicates that the pregnancy rate in New York City did not increase after the 1965 blackout according to the information posted at www.snopes.com/pregnant/blackout.asp.

both blackouts are remembered primarily for their effect on New York City, the outages themselves resulted from events that transpired hundreds of miles from the city. Those two outages highlight the complex interconnected nature of our electrical power grid, which is discussed in detail in Chapter 7. Interestingly, the New York City blackout in 1977 originated with a mechanical failure in the city and affected only the city. Unlike the 1965 and 2003 blackouts, the 1977 blackout did cause prompt significant unrest.

Political discussions about energy often divide energy sources into three broad categories: fossil fuels, renewables, and nuclear systems. Figure 2.4 graphs the average annual electric power generation in the United States, where the sources are grouped into these three major categories. Interestingly, for all of the changes in primary energy sources discussed above, the fraction of U.S. electrical power originating from nuclear, renewables, and fossil fuels has remained remarkably stable since 1980 even though the total consumption has increased by approximately a factor of two. Figure 2.4 shows that since around 1980, 70 percent of U.S. electrical power was provided by fossil fuels, 20 percent was provided by nuclear power, and 10 percent was provided by renewables.

From a scientific perspective, it is useful to divide these three major categories to reflect the actual primary energy sources, as shown in Figure 2.5. This figure shows that coal has long been dominant but that its role is rapidly decreasing as natural gas use increases. Natural gas use is increasing predominantly because fracking has increased the availability enough to result in dramatic price reductions, though the lower emissions associated with natural gas burning are probably also playing a role. The fractional and absolute power provided by petroleum showed a sudden decrease after the oil shock of the late 1970s. That decrease was followed by a rather stable plateau, which has since been followed by a steady decrease since 2005. At present, petroleum makes little or no contribution to U.S. electrical power generation even though it is the dominant energy source for transportation.

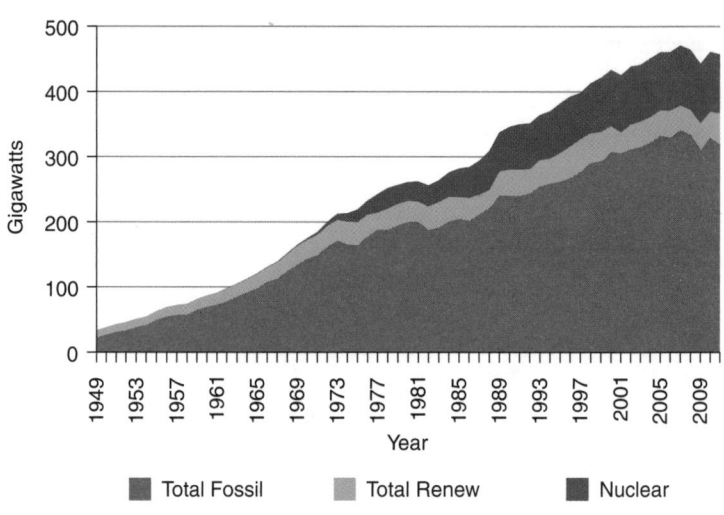

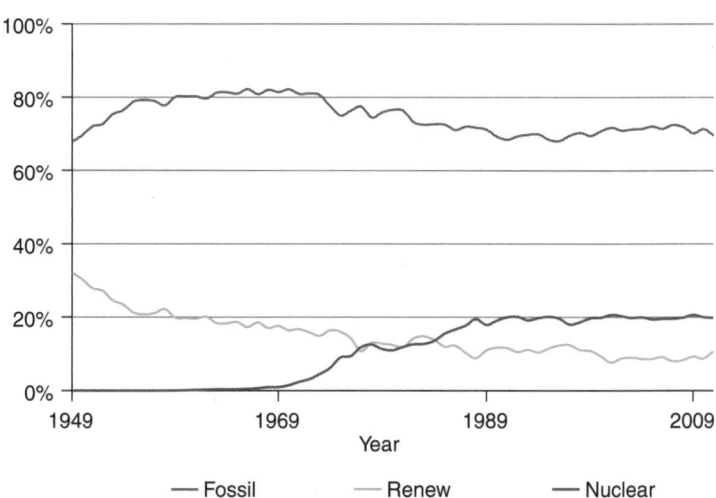

Figure 2.4 Contributions of fossil fuels, nuclear energy, and renewables to the total electric power generation in the United States. The top graph shows the actual contributions, and the bottom graph shows the percentage contributions. The nuclear contribution has been stable since around 1990, as has the total renewable contribution. The stability of the fractional contribution of renewables masks a decrease in the hydroelectric contribution, which has been stabilized by the increase in other renewables. Current and past data is available at www.eia.gov/forecasts/aeo/er/index.cfm in topic A8.

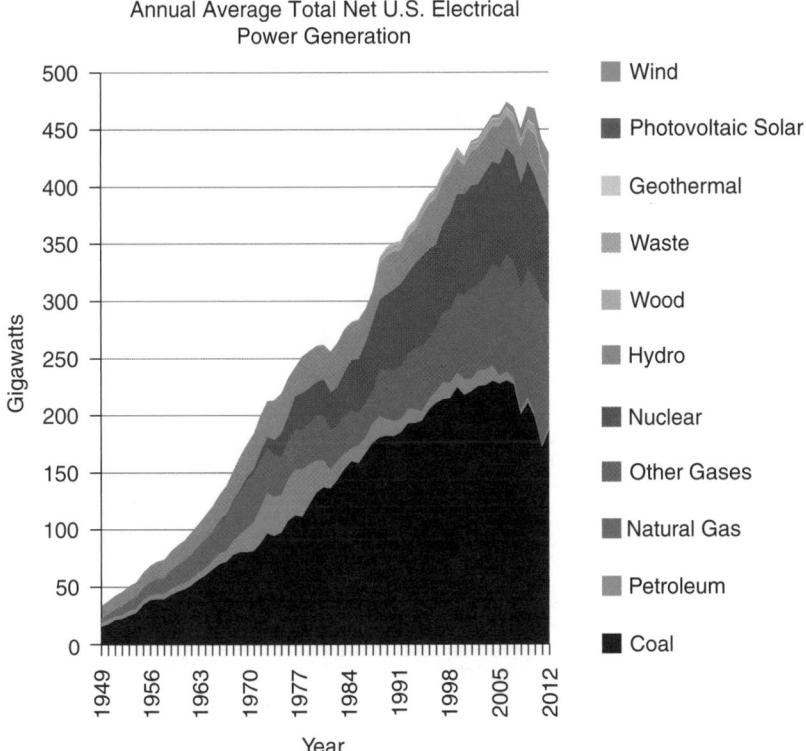

Figure 2.5 Average annual total U.S. electrical power generation, showing a steady increase since 1949 until around 2005 and then a leveling followed by a slight decrease since.

Not only have the primary sources of fossil fuel energy evolved over time, but Figure 2.6 shows that the composition of the renewables has changed significantly over time, even though Figure 2.4 shows that the total fractional contribution of renewables has remained stable. Figure 2.6 shows that the power generated by hydroelectricity has remained constant since the mid 1970s. As I will discuss later in this chapter, the United States is near the practical limit for hydroelectric power generation, so the total power generated from hydroelectric is unlikely to ever significantly increase. Though the power generated by hydroelectricity has

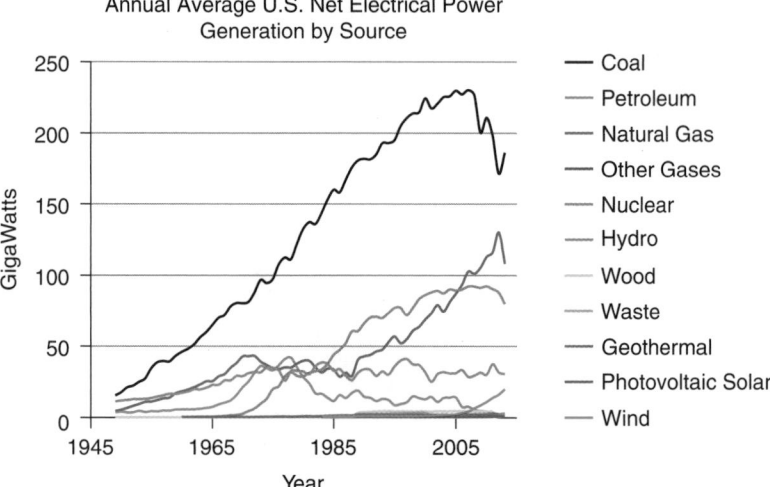

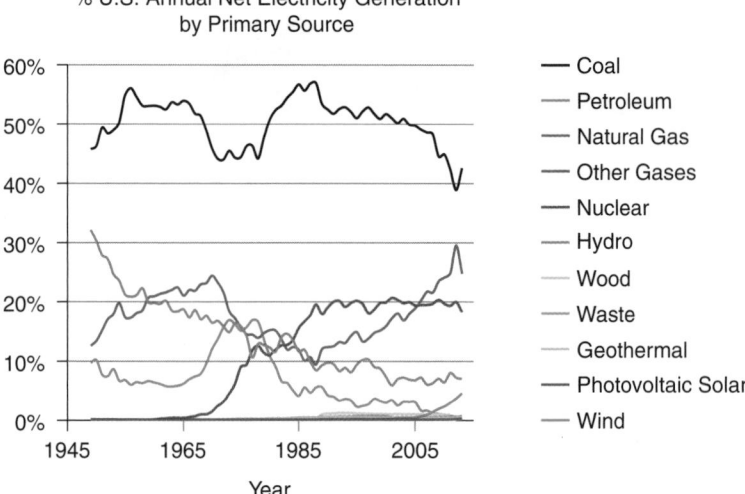

Figure 2.6 Annual average U.S. electrical power generation by source. The upper left graph shows the actual power, and the lower left graph shows the fractional distribution of the power among the sources. The graphs show that coal continues to dominate, while natural gas has been steadily increasing. The absolute contribution of hydropower has been flat since around 1975, so the increase in total generated power has resulted in a large decrease in the fractional contribution of hydropower. The slight absolute increase in nuclear since around 1985 has allowed the fractional contribution to remain stable at 20 percent. Petroleum picked up pace in the early 1970s but decreased after the oil shock and has been disappearing since. The graphs on the right show

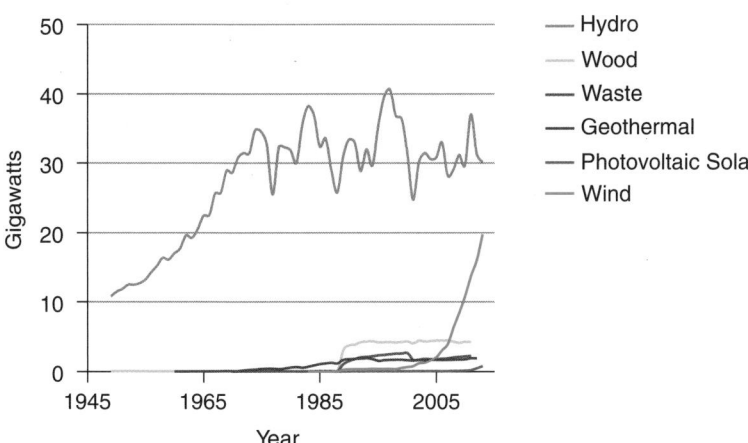

Annual Average U.S. Net Electrical Power Generation by Renewable Source

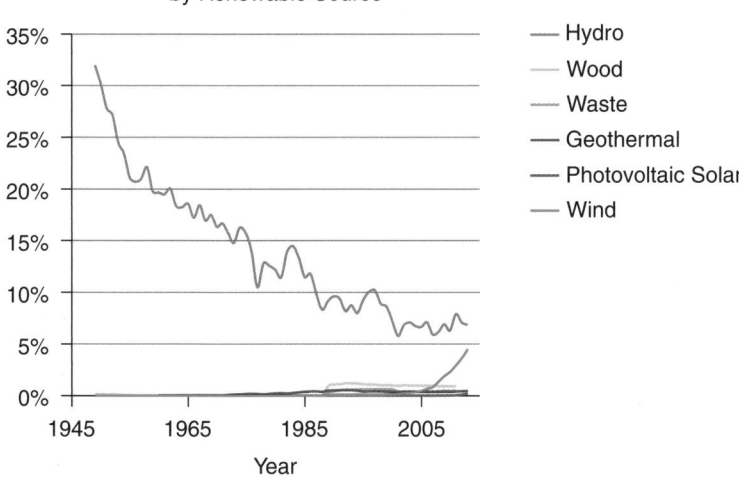

% U.S. Annual Net Electricity Generation by Renewable Source

only the annual average electric power contribution of renewable sources. Actual consumption is shown in the upper graph, and percentage contribution is shown in the lower graph. Wind has been increasing dramatically since 2005 and now provides more than 4 percent of U.S. electrical power, whereas geothermal has been fairly stable. Photovoltaic solar is too small to see even on this scale. Large discontinuous jumps between years probably represent changes in how some sources are categorized rather than real changes. Available at www.eia.gov/forecasts/aeo/excel/aeotab_8.xlsx.

remained constant since around 1975, U.S. electric power use has dramatically increased. As a result, Figure 2.6 shows that the percentage generated by hydroelectric power is decreasing. This trend will probably continue. In contrast, the fraction provided by nuclear power has remained approximately constant because the increase in nuclear power generation has roughly kept pace with the increase in electrical power consumption. Though nuclear power is nowhere near a practical limit, it may be near its economic and political limit. In Figure 2.6, it is difficult to see the contributions of wind and solar PV because they are still so small. In order to see the contributions of wind and solar more clearly, Figure 2.7 shows only the contributions of non-hydroelectric renewables.

Figure 2.7 shows that since 2005 there has been a strong increase in wind power. Solar power has shown a strong increase since around 2010, but it is important to note that these are graphs of net generation. If electricity is generated and used without being fed into the grid, then that contribution does not appear. It is likely that most of the power generated by rooftop solar panels is used directly by the owners of the panel and not fed into the grid, which explains why the net electrical contribution of solar power appears to be smaller than the total energy contribution of solar that was considered in the Introduction.

I will argue that even given existing technology, wind and solar PV systems could greatly increase their absolute and fractional energy contributions. Hydroelectric power's inability to increase in absolute terms will freeze its fractional contribution if electricity use remains level or result in continued increases if electricity use begins to increase again; however, if economics and politics keep nuclear capacity from increasing significantly while wind and solar continue to increase rapidly, in the United States the electricity provided by renewables should exceed the contribution of nuclear in the next ten years despite the fixed contribution from hydroelectric power.[10]

10. As shown in Figure Intro.4, renewables already contribute more total energy than does nuclear power as a result of the significant contribution of biofuels that are not used to generate electricity.

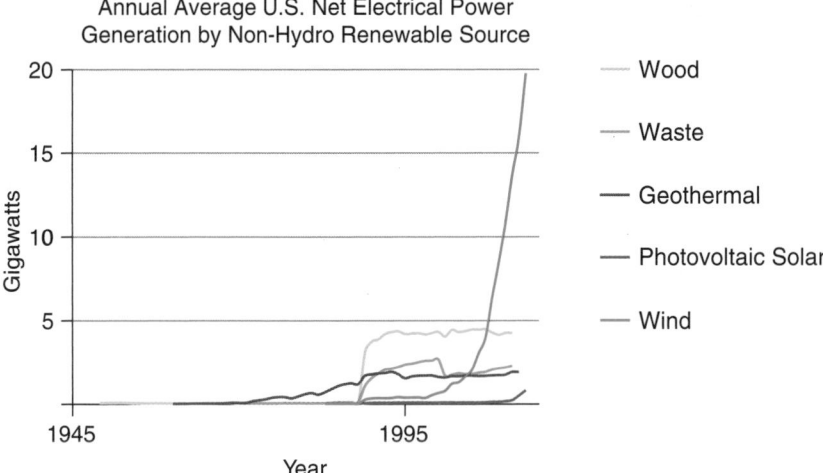

Figure 2.7 Annual average electric power contribution of non-hydro renewable sources. The dramatic increase in windpower since 2005 is clearly visible. Photovoltaic solar remains less than 0.5 percent of the net contribution, but it has been increasing rapidly since around 2010. Large discontinuous jumps between years probably represent changes in how some sources are categorized rather than real changes.

One question this book cannot answer is whether U.S. electricity use has "permanently" leveled off or whether it will start to rise again. If it continues to rise, increases in renewables can be used to meet increasing demand without decreasing demand from existing plants. Thus, the additional cost of switching to renewables would include only the difference between new plants using fossil fuel or nuclear fuel and new plants using renewables. If U.S. electricity use remains constant, then demand for new generating capacity will be small, so increases in renewables will result in decreases in use from other sources. Thus, existing plants will deliver lower than expected returns on capital, so the cost of switching to renewables must include not only the cost of the new physical plant but the decreased use of the existing physical plant. In contrast, if the United States switched from petroleum-powered vehicles to electrically powered vehicles, the amount of electricity required could drastically increase, even if overall energy consumption was substantially reduced because of increases in efficiency. This demand for new electricity sources could be met by new renewable sources without requiring the retirement of any fossil fuel–burning plants, so the cost of the new renewables would not need to include that additional factor.

In this book, I will present some cost comparisons based on building new physical plants using various primary energy source to determine whether particular options are economically ridiculous; however, I caution readers to remember that many other factors may affect the overall cost of changes in energy consumption, including subtle effects like increases in food prices that result when more cropland is devoted to biofuel cultivation.

OVERVIEW OF MECHANICAL TO ELECTRICAL ENERGY CONVERSION

With the exception of solar PV systems, electrical generators use mechanical work to rotate a shaft. The rotating shaft is attached to a system of wire loops and magnets that causes electrons to move down the wire as the shaft is rotating. In hydroelectric systems, falling water induces shaft rotation; in combustion systems, burning fuel produces the energy used to rotate the shaft; in solar thermal systems, the heat generated by sunlight provides the energy to rotate the shaft; and in wind and wave power, the motion of the air or the water causes the shaft to rotate. In this section, I will examine how a rotating shaft can cause electrons to move along a wire and provide electrical power that can be used by residential, commerical, and industrial consumers. In Chapters 3, 4, and 5 I will consider how water, wind, and combustion are used to rotate a shaft.

FARADAY'S LAW

Early electrical generators made prior to 1831 were made without an understanding of how they worked, but in 1831–1832 Michael Faraday discovered and demonstrated how electrical generators actually worked. The basic physics that underlies generators is one of the most magical features of physics. Somehow rotating a wire loop located near a magnet make electrons move around the wire loop, even though the magnet and the wire never touch. This interaction is an example of action at a distance. For people a hundred years ago, the concept was very strange; for some it was even creepy. However, in the age of wireless communication, someone in Tokyo can use an app on her smart phone to turn down the heat in her apartment in Boston. Thus, the idea of action at a distance seems much less odd than it once did. In the particular case of Faraday's experiments, the mysterious action at a distance was summarized in Faraday's law, which is discussed in detail in the online Appendix 5 at http://thedata.harvard.edu/dvn/dv/HUP.[11]

Faraday's law would have warmed the heart of Heraclitus, the famous Greek philosopher who believed that it is only change that matters. What Faraday found out through experimentation is that when a moving magnet passes across a wire loop, a lightbulb attached to the loop will light just as the magnet is passing across the coil. If the magnet passes over the loop repeatedly, the lightbulb will light while the magnet is moving across the loop, but the bulb will not light if the magnet is held steadily over the loop. Thus, the lightbulb lights when the magnet is changing its position with respect to the wire loop, in accordance with Heraclitus's suggestion that only change matters.[12] In sum, for electrical generators

11. According to Wikipedia, Faraday had no formal scientific education, though he was one of the most gifted experimentalists of all time. Albert Einstein kept a picture of Faraday on the wall of his study along with pictures of Isaac Newton and James Clerk Maxwell. Though physicists think of Faraday as a physicist, chemists prefer to think of him as a chemist. As a chemist, Faraday pioneered electrolysis, which is the process by which electrical current can separate the hydrogen and oxygen in water to produce gaseous hydrogen and oxygen. This practical invention may eventually provide the hydrogen for fuel cell cars. See en.wikipedia.org/wiki/Michael _Faraday, accessed June 22, 2014.

12. In addition to explaining how to generate electricity, Faraday's law also provides the basis for describing electrical transformers, which can transfer electricity between two wires that are not physically connected, resulting in a transfer of voltage that can

Faraday's law can be summarized as follows: a moving magnet can make electrons in nearby wire loops flow, providing a source of electrical power.

Bicycle odometers use precisely this idea to count the number of times that the wheel rotates per minute. Bicycle odometers consists of a permanent magnet attached to a spoke on the wheel and a wire loop attached to the fork of the bicycle to which the hub of the wheel is attached. Thus, the rotation of the bicycle wheel constantly changes the relative distance between the magnet and the wire loop. This change generates electricity. In the bicycle odometer, each time the magnet passes the wire loop, a pulse of moving electrons is sent from the loop to the computer on the handlebars of the bicycle. The computer counts the number of pulses per second. If you have correctly programmed the computer with the diameter of the wheel of your bike, then with the aid of its internal clock, the computer will tell you your current speed, your peak speed, and your distance traveled.

The electrical pulses in the bicycle odometer were the result of pulses of moving electrons. Moving electrons have kinetic energy, so conservation of energy (see online Appendix 3.4) requires that energy must come from somewhere in order to generate the kinetic energy in the moving electrons. In the bicycle odometer, the energy came from the rotational kinetic energy of the rotating tire. In electrical power plants, the energy comes from the mechanical work (see online Appendixes 2 and 3) that

---

be higher than, lower than, or the same as the initial voltage. The power delivered to a device is a product of the current I and the voltage V. Thus, power = IV. As mentioned previously, in SI units power is measured in watts, current is measured in amps, and voltage is measure in volts. One might first hope that when a transformer increases the voltage, the delivered energy would increase; however, energy can neither be lost nor gained. Thus, the higher voltage is always associated with a lower current. An ideal transformer would conserve energy, so the output power would be the same as the input power; however, real transformers actually lose energy when they convert voltage. In fact, much of the electrical energy that is "lost" during electricity delivery in the United States today is lost to exactly those power transformers.

is required to rotate a shaft so that the relative position between a magnet and a wire loop is constantly changing. Unending work is required to maintain a continuous change in relative position. This continuous work provides the energy that electricity companies deliver to consumers. An extremely detailed discussion of how mechanical energy is transformed into electrical energy is located in online Appendix 5, but a short general discussion of the principles involved will be presented here in terms of the Greek myth of Sisyphus. Both cases require a never ending push in order to provide the mechanical work necessary to constantly move an object that resists being moved.

In the Greek myth of Sisyphus, Sisyphus suffered eternal punishment. He was required to push a boulder up a hill, but when the boulder reached the top, the boulder rolled back down to the bottom and Sisyphus had to push it back up again.

Various interpretations of the myth of Sisyphus have been proposed. In 1942, Albert Camus suggested that the myth of Sisyphus might be interpreted as a source of joy rather than torment if one embraces the "hopeless" task and defines success as doing the task rather than completing it, or as the modern bumper sticker says, "It is the journey, not the destination." The myth of Sisyphus was revisited in the 1980s when atomic physicists were trying to understand how light could cool atoms to unimaginably low temperatures, well below what people believed to be the fundamental quantum limit. The answer turned out to be described by the "Sisyphus effect," in which the atom moves through the hills created by a reflected light wave. The interaction between the light and the atom is such that every time the atom gets to the top of the hill, the interaction changes and what was the top of a hill suddenly becomes the bottom.

In an electrical generator, the act of mechanically rotating the wire loop near a permanent magnet induces electrons to move around the loop. Those moving electrons are delivered to consumers as electrical power. When the electrons are moving in the loop, they turn the loop into a "virtual magnet." It is well known that if one tries to push two magnets together, they will resist the push if the magnets are oriented such that like poles are facing each other. In contrast, if opposite poles are facing, then work is required to pull the magnets farther apart. In an electrical generator, the poles in the "virtual magnet" are always oriented so that the rotation of the loop is opposed by the interaction between the real permanent magnet and the "virtual magnet" that is created by the electricity generated by the moving magnet.[13] Thus, the electricity generated by the shaft rotation results in a force that always opposes the shaft rotation, so like Sisyphus, the power generator is doomed to forever push uphill; however, unlike Sisyphus, who is simply being tortured eternally, the endless shaft rotation produces a positive result: the work done to rotate the shaft results in the delivery of electricity to consumers. A detailed discussion of how Faraday's law results in the generation of electrical power is presented in online Appendix 5.

Given that a turning shaft can generate electricity, we will now consider how different energy sources can be used to make a shaft rotate. Thus, we will consider the conversion of potential, kinetic, or heat energy to the mechanical work that is required to turn a shaft.

13. In other words, the work done to rotate the loop creates a virtual magnet that interacts with the real permanent magnet, resulting in a push that resists the rotation of the loop and the shaft to which it is attached.

RENEWABLES ARE ENOUGH II

3

ELECTRICITY FROM WATER

In hydroelectric systems, falling water does the mechanical work required to rotate a shaft and generate electricity. Hydroelectric power is an exceptional source of renewable energy because it provides built-in energy accumulation and storage, allowing hydroelectric power systems to respond easily to changes in electricity demand despite the intermittent nature of rainfall. Early in the twentieth century, dams were considered amazing human achievements. The recent completion of the monumental Three Gorges Dam in China is a source of great pride to many; however, as the environmental, cultural, and social impacts of large dams have become clear, people have become less enamored of dams. As will be discussed in Chapter 10, in the United States, some functioning hydroelectric dams are actually being removed in order to improve the environment. Figure 1.1 shows, hydroelectric power generation in the United States has been approximately constant since around 1970. Given that electricity use has been increasing, the fraction of electricity provided by hydroelectric systems has been steadily decreasing. At present, hydroelectric power provides around 30 gigawatts of electric power, which is only about 7 percent of the electric power used in the United States. The U.S. Department of the Interior estimates that there is around 30 gigawatts of undeveloped hydroelectric power in the

United States;[1] however, the department acknowledges that social and economic factors suggest that much of that "undeveloped" power will never be used. Even if all of that power was developed, hydroelectric power would provide less than 15 percent of the electric power now consumed in the United States. Furthermore, in this chapter I will show that the amount of rainfall in the United States is insufficient to provide all of the electrical power needs even if all of the rainfall in the United States could be captured and used.

As it currently stands, we are already exploiting a substantial fraction of the rain that falls on the United States. It is possible that increases in hydroelectric power in the future will be associated with energy storage rather than energy harvesting. The water stored behind a dam allows hydroelectric system to provide energy on demand, where the amount of power provided can be changed almost instantaneously. Thus, hydropower can at least partially compensate for variations in wind and solar power without requiring the burning of fossil fuels. If an intermittent energy source, such as solar power or wind power, is used to pump water uphill during times when electrical supply exceeds demand, then the stored water can later be allowed to fall back downhill and generate electricity when the demand exceeds the supply from solar and wind sources. The storage volume need only be sufficient to cover the times during which the intermittent power source is inadequate. For the example of solar power, this largely means storing energy during the day to be used during the night. If the elevated water is stored within a lake, then the system simply pumps lake water into and out of a raised reservoir within the lake, so available rainfall would not be an issue. Such systems will be discussed in detail in Chapter 9, which considers energy storage. In this chapter, I will consider only how the energy stored in elevated water

1. The undeveloped power is around 19 gigwatts in Oregon, 10 gigawatts in Pennsylvania, and 7 gigawatts in South Dakota, while remaining states offer less than 5 gigawatts each. See www.usbr.gov/power/edu/pamphlet.pdf.

is converted into the mechanical energy of a rotating shaft that generates electricity.

HOW HYDROELECTRIC DAMS WORK

Flowing water has long been used to make shafts rotate. In the 1700s and early 1800s, the rotating shafts were directly attached to grinding wheels in mills and machines in factories. Richard Arkwright played a significant role in the Industrial Revolution. He invented a spinning machine that was more effective than machines used previously, but the machine required more mechanical work than the previous machines, which were turned by people. Arkwright used a water wheel to provide the required energy. The new machines could also be operated by less skilled workers, including children. Arkwright himself refused to employ anyone younger than six years old, but others were less scrupulous. Arkwright's factory was so successful that the local population was insufficient to provide the necessary workers. He solved the problem by building a new village and importing workers from the countryside to live in the village and work in the factory.

Mills and machines are now driven by electrical power rather than falling water, but some of that electrical power is still generated by falling water. In hydroelectric power plants, flowing water turns a turbine that rotates a wire coil in an electric generator. A simplified illustration of the complete energy cycle for a hydroelectric power system is shown in Figure 3.1. The sun delivers heat energy to the earth. Some of that energy heats water on the surface of the earth, and the heated water evaporates and rises up. The heat energy delivered by the sun is converted to gravitational potential energy when the heated vapor rises. Thus, hydroelectric power originating from energy provided by sunlight is really a form of "solar power" even though we don't generally think of it as such.

The elevated water in clouds has a gravitational potential energy. If water rains down on top of a mountain that is h meters tall, then at the

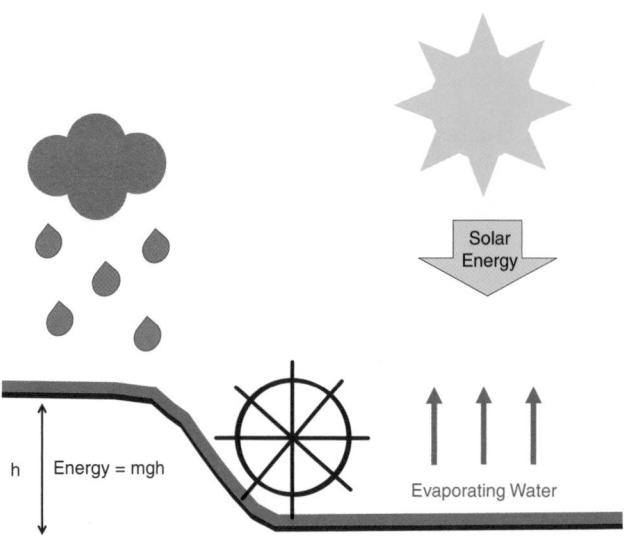

Figure 3.1 Hydroelectric power cycle showing that the sun provides the energy that evaporates water at the surface of the earth. The evaporated water rises and forms clouds that are high above the earth. As a result, the water in the clouds has a large gravitational potential energy. When the water rains down from the clouds, the water is deposited in elevated regions. That water collects and flows toward large bodies of water, including the ocean. The sun then evaporates the water again, and the cycle can continue as long as the sun shines.

top of the mountain the water has a gravitational potential energy of mgh, where m is the mass of the water, and g is little g, as discussed in online Appendixes 2 and 3 at http://thedata.harvard.edu/dvn/dv/HUP. As the water falls down the mountain because of gravity, the kinetic energy of the falling water can turn a water wheel. The turning of the water wheel produces a rotation in the shaft of the electrical generator. Faraday's law allows the rotation of the shaft to create a flow of electrons that moves along wires to customers who consume the electrical energy.

Falling water can lose energy to friction as it flows along a stream; therefore, to avoid this energy loss, people have employed dams to force the water to fall down a large height without interacting with the surface of the earth. People find falling water very pretty—pretty enough to make Niagara Falls a honeymoon destination and Angel Falls in Venezuela a tourist gold mine. However, modern hydroelectric plants generally do not produce dramatic displays of falling water. Exceptions occur during spring flooding, when hydroelectric dams sometimes release large amounts of water, producing conditions that allow white water kayaking and rafting on what is otherwise a slow moving river.

During the regular operation of most modern hydroelectric power plants, gravitational potential energy is not converted to kinetic energy by allowing the water to freely fall from a height. Rather, water is released at the bottom of the dam. At first, this would seem to be quite silly, but such systems are actually very efficient because they avoid frictional losses associated with the water falling a long distance.

A fairly realistic diagram of a modern hydroelectric power system is shown in Figure 3.2. It consists of a large dam, which stores water that flowed downhill to the dam after raining down from the clouds; however, as discussed above, the water does not flow over the top of the dam. Instead, the water that runs the electrical generator flows out of a controllable intake gate near the bottom of the dam. After passing through the gate, the water passes into the penstock, which is a tunnel that transfers water from the dam to the turbine. The rate at which the water flows is controlled by the position of the intake gate. The water flows through the penstock because of the large weight of the water resting above it in the dam. Water or wine that comes packaged in a box with a spigot at the bottom uses the same principle: the weight of the liquid inside the box exerts a force that pushes the liquid out through the spigot at the bottom of the box, where the flow rate decreases when there is little liquid left in the container because the weight of the remaining liquid is so small that it exerts very little force.

Similarly, in a hydroelectric dam, the water at the bottom of the dam must exert a force to "hold up" the weight of water on top of it, just as the mythological figure of Atlas exerted a large force to hold up the earth. As discussed in online Appendix 2, pressure is simply defined as the force over a unit area. Thus, the force associated with holding up the water can

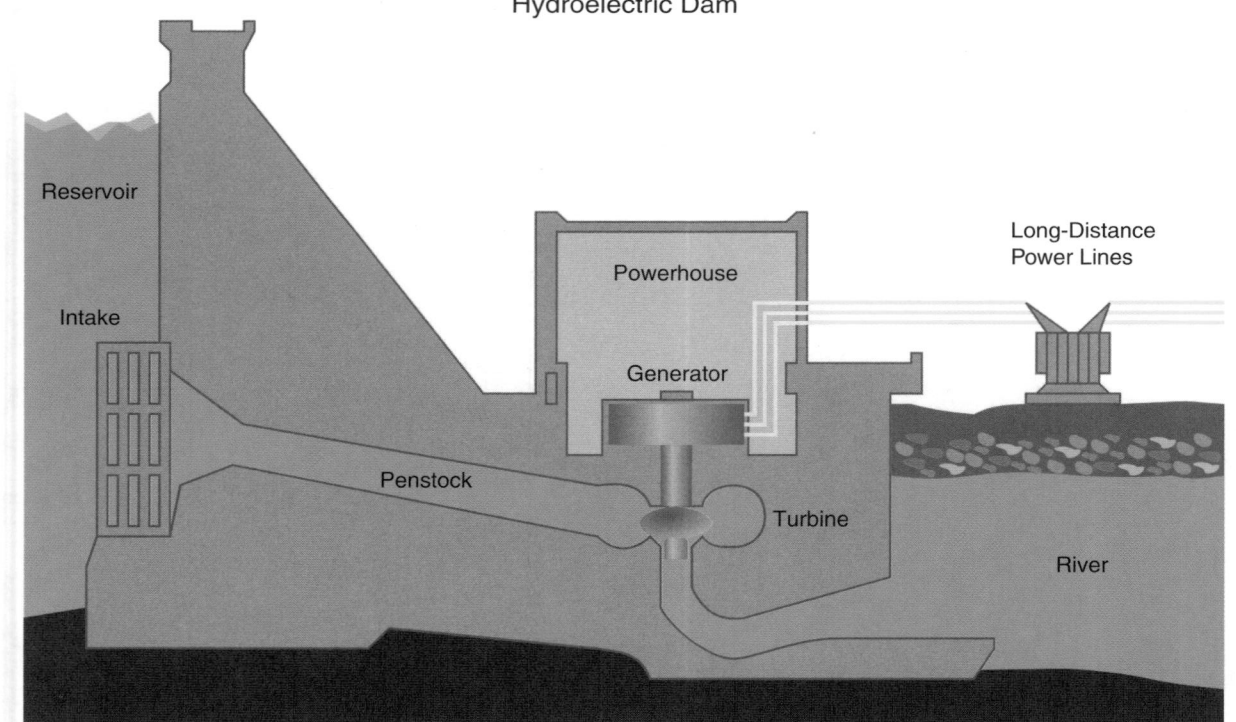

Hydroelectric Dam

Reservoir

Intake

Penstock

Powerhouse

Generator

Turbine

Long-Distance
Power Lines

River

Figure 3.2 Schematic of a hydroelectric power plant showing the flow of the water from behind the dam. The water flows through the intake gate and down the penstock before pushing the turbine in order to rotate the generator shaft and produce electricity that can be transmitted to consumers. This image is from the work of Tomia based on www.tva.gov/power/hydroart.htm.

be discussed in terms of the pressure that the water in the dam exerts on the water at the bottom of the dam. At the top of the dam, the water holds up only the air above it, so the pressure is equal to the pressure of the air, which is usually called atmospheric pressure. Water at lower levels must hold up the water above it as well as the air, so the water pressure increases as one moves from the top of the dam to the bottom of the dam.

The pressure that one feels in one's ears when swimming from the top of a swimming pool to the bottom provides a visceral measure of the increase in pressure as the water depth increases. One can calculate the pressure exerted by the water in the swimming pool by calculating the weight of the water sitting above the bottom of the pool and then dividing by the area of the bottom of the pool to get the pressure, as illustrated in Figure 3.3.

In mathematical terms:

Water force on the bottom = Water weight in the pool
$$= g \times \text{(mass of the water in the pool)}$$
$$= g \times \text{(the density of the water} \times \text{the volume of the water in the pool)}$$
$$= g \times (\rho \times \text{volume})$$
$$= g \times (\rho \times h \times a \times \text{area}).$$

Thus, the pressure = force per unit area = force/area is equal to the product of the gravitational constant, the density of water, and the height of the dam. In scientific writing the $\times$ sign is often left out; therefore, pressure = force/area = $g \times \rho \times h$ is usually written as $g\,\rho\,h$. Thus, in accord with our intuition, higher dams provide higher pressure. Notice that the volume of the water in the dam does not affect the pressure. Only the height matters.

Even though the pressure is not affected by the total volume of the water behind the dam, the total energy stored behind the dam is proportional to the amount of water stored. If instead of depending on rainfall to deliver water into the reservoir behind the dam, one uses electricity to pump water into the reservoir, then electrical energy is converted into stored gravitational energy.

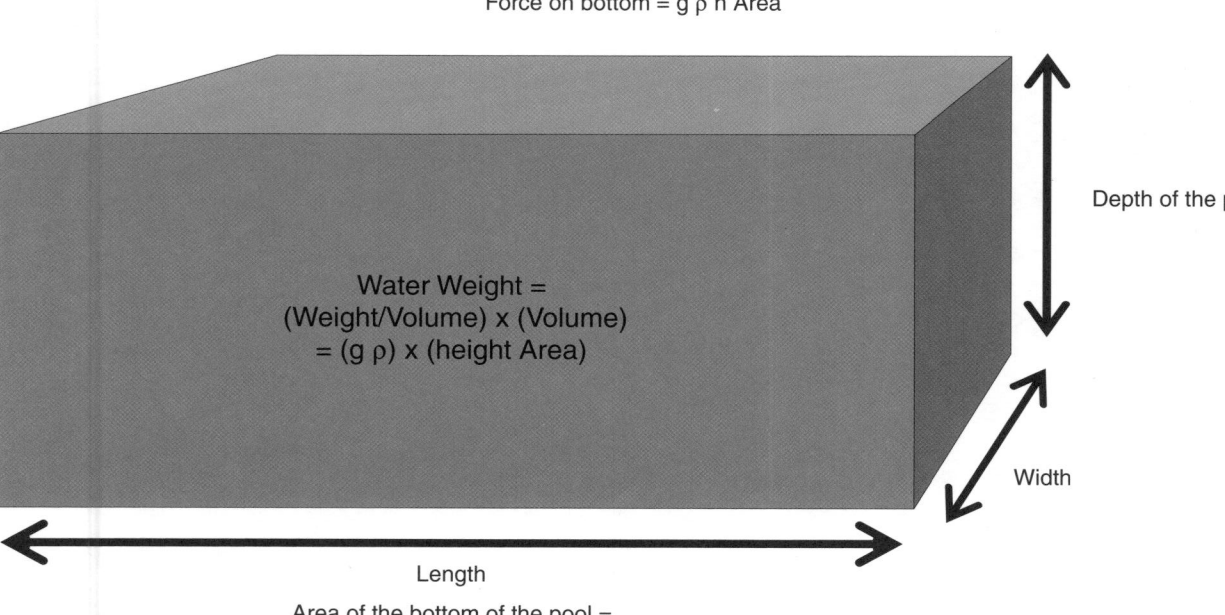

Force on bottom = g ρ h Area

Depth of the pool = h

Water Weight =
(Weight/Volume) x (Volume)
= (g ρ) x (height Area)

Width

Length

Area of the bottom of the pool =
Length x Width

Figure 3.3 Illustration of how to calculate the force exerted on the bottom of a pool by the water in the pool. The bottom holds up the water, so the force on the bottom is equal to the weight of the water. The weight of the water is equal to the mass of the water $\times$ g. The mass of the water is equal to the water mass per unit volume $\times$ volume of the pool $= \rho \times$ Volume $= \rho \times h \times$ Area. Thus, the force on the bottom is equal to $g \times$ mass $= g \times \rho \times h \times$ Area.

Different hydroelectric systems translate the pressure exerted by the water into shaft rotation in different ways. There are two main types of generators: impulse generators and reaction generators. In both cases, the high pressure of the water is converted into kinetic energy by allowing the water to fly out through a narrow nozzle, where narrower nozzles produce higher water velocities. This effect appears when someone puts his thumb partially over the end of a garden hose. For most people, it is not possible to fully block the end of a hose with a thumb because the pressure exerted by the water is too great. Thus, the total amount of water flow is usually not changed much by the presence of the thumb, but the water flies out much faster and travels much farther as the area blocked by the thumb increases.

In an impulse turbine, the nozzle is stationary. The water flowing out of the nozzle hits the blades of the turbine and pushes back on them, which causes the shaft attached to the blades to rotate. This system resembles the traditional water wheel: moving water pushes on the blades and makes the blades rotated.

In the reaction turbine, the nozzle is attached to the shaft, but the nozzle is pointed in the direction of rotation. This system is analogous to spinning garden sprinklers. The water flying out of the nozzles pushes the arms, making them rotate about the central shaft. The same effect causes a balloon to fly forward when air is flowing out the back of the balloon. In technical terms, both effects are a result of conservation of momentum, which is discussed in detail in online Appendix 3.4.

CALCULATION OF THE POWER GENERATED

It is possible to calculate the energy transfer in such systems by performing detailed calculations of the velocity of the water at each point in the system, but it is much easier to use the magic of conservation of energy (see online Appendix 3.4). Figure 3.4 shows a simplified version of a hydroelectric system into which no water is flowing.

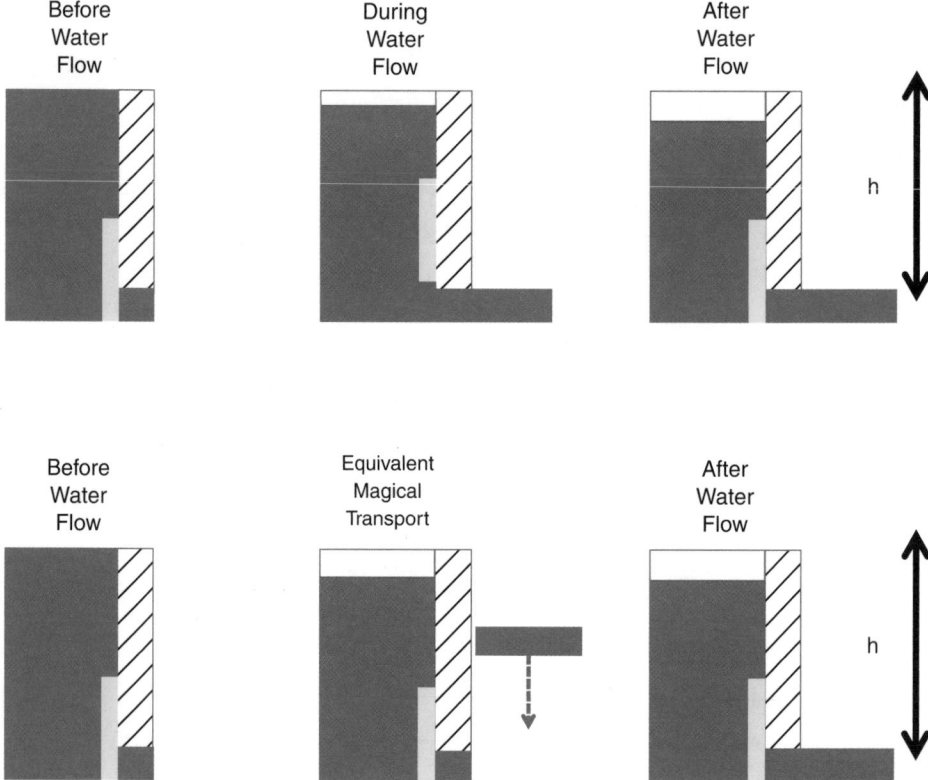

Figure 3.4 Idealized schematic to simplify the calculation of the maximum energy that could be obtained from a hydroelectric dam. The upper panels illustrate actual flow in a system where no water is coming into the dam. The water is shown in blue, and the dam is indicated by the rectangle containing black diagonal stripes. Initially, the intake gate (yellow rectangle) to the penstock is closed, and no water is flowing. When the intake gate opens, water flows through the penstock and into the river below the dam. After some time, the penstock is closed again. As a result of the water flow, the water behind the dam has a lower height than before, but there is more water in the river at the foot of the dam, as indicated by the white rectangle at the top of the dam and the blue region extending to the right at the bottom of the dam. The details of the water flow may be complex, so calculating the energy transfer from the water to the turbine can be very difficult. The lower diagram is useful because it allows one to calculate the maximum possible amount of energy that could be delivered to the turbine as a result of the flowing water if all of the gravitational potential energy in the water were delivered to the turbine.

The blue indicates water, and the black and white stripped region indicates the cement in the dam holding back the water. The top of the water is a height h above the output river. At a time $t=0$, water is allowed to flow through a penstock to a turbine generator. After flowing through the turbine, the water is released into the river. The actual path is very complicated, and in general the water will leave the dam with some kinetic energy; however, the maximum energy that the water can contribute to the turbine is the change in the potential energy between a volume of still water with mass m at the top of the dam and the same still water at the bottom of the dam. This is the situation that would result if the water were enclosed in a weightless container that was transported very slowly to the bottom of the dam where it was gently deposited at the output. That change in energy is very easy to calculate: it is simply the change in the gravitational potential energy of the water, which is $m \times g \times h$, where m is the water mass in kilograms and h is the height in meters. Power is the change in energy over time, so the power delivered is equal to the amount of mass that flows through per second times equal to the product of g times the height $(g \times h)$. All systems have some losses, which are not included in that equation. Losses are usually included by multiplying the ideal energy change by an efficiency factor. Modern hydroelectric systems have efficiencies of approximately 90 percent,[2] so that factor does not change the result much.

Power in watts =
$$= \text{efficiency} \times (\text{mass of water flowing through in one second}) \times g \times h$$
$$= \text{efficiency} \times [\text{mass/(liter of water)} \times \text{liters/second of water flow}]$$
$$\times 9.8 \times h$$
$$= \text{efficiency} \times (1 \text{ kg/liter} \times \text{liters/second}) \times 9.8 \times h$$
$$= \text{efficiency} \times \text{liters/second} \times 9.8 \times h$$
$$\sim \text{liters/second} \times 9.8 \times h$$

2. See www.usbr.gov/power/edu/pamphlet.pdf.

Thus, the power generated by a hydroelectric dam increases linearly with the height of the dam and the water flow rate.

Even the largest source of hydroelectric power generation in the world, Three Gorges Dam in China, has a nominal capacity of only 20 gigawatts, which is 20 billion watts or 20 million kilowatts. The entire power of even that dam would be enough to support only 20 million American households. For comparison, in the United States, the Grand Coulee Dam generates around 7 billion watts and the Hoover Dam only 4 billion watts, so together they provide about 3 percent of U.S. electrical power, which is almost half of the total hydroelectric power generated in the United States. In contrast, the Aswan Dam in Egypt provides around 15 percent of Egypt's total power, even though it has a nominal capacity of only 2.1 gigawatts. Table 7 shows the power capacity of major dams. All of these projects required large capital investments and topographical changes, and some also involved significant displacement of populations. For example, the Three Gorges Dam required the relocation of more than one million people. In addition, uncounted numbers of animals lost their habitats.

Of course, hydroelectric power is unsustainable if average water outflow from the reservoir exceeds the average water inflow, so for a system that depends purely on rainfall to fill the reservoir, the average amount of water flow used to generate energy cannot exceed the total rainfall that is directed to the dam by rivers and springs that provide water to the res-

Table 7.   Information on the five largest hydroelectric power dams in the world.

Rank	Dam	Country	Capacity (MW)	Number of U.S. residences that could be served	Fraction of U.S. electricity us (in percent)
1	Three Gorges	China	20,300	16,000,000	4.6
2	Itaipu	Brazil/Paraguay	14,000	11,000,000	3.2
3	Guri	Venezuela	10,200	8,000,000	2.3
4	Tucurui	Brazil	8,370	6,000,000	1.9
5	Grand Coulee	United States	6,809	5,000,000	1.5

Source: http://en.wikipedia.org/wiki/Hydroelectricity.

Table 8.   Estimated maximum possible hydroelectric power generation by the United States, compared with the nation's current energy needs. The table shows that hydroelectricity could not even meet the current U.S. electrical needs.

Quantity characteristic of hydroelectric power in the United States	Value
Average U.S. rainfall per year	0.7 m
Energy/m^2/year, assuming 100-meter dam = mgh = water density × water volume × 9.8 × 100 = 1,000 kg/m3 × (0.7 × 1 × 1) × 9.8100	686,000
Watts/m^2 [energy/sec/m^2 = energy/year/(sec/year)]	0.022
U.S. surface area (m^2), excluding water	9.E+12
Total possible energy for 100% of U.S. water collection	2.E+11
Watts/m^2 for total U.S. energy consumption	0.33
Percentage of total U.S. energy, given 100% possible hydro	7%
Watts/m^2 for energy used, after 50% losses, which are largely thermal	0.16
Percentage of total U.S. energy after losses, g 100% hydro	13.35%
Watts/m^2 for total U.S. electric consumption	0.04
Percentage of total U.S. electrical energy used, given 100% hydro	48%

ervoir. Thus, the amount of usable rainfall limits the power that can be obtained from hydroelectric systems.

HYDROPOWER CANNOT MEET U.S. ELECTRICITY DEMAND

According to the World Bank, the average annual rainfall in the United States is approximately 0.7 meters a year.[3] As a result, the total mass of the water delivered to one square meter in a year is approximately 700 kilograms, so the average rainfall per second is 2×10^{-5} kilograms. If that water were harnessed behind a 100-meter-tall dam, the average energy associated with that flow would be the product of the flow per second times the gravitational constant g times the height of the dam = 0.02 watts per square meter.[4] Thus, if the system were 100 percent efficient, generating the one kilowatt used by the average American household would

3. See http://data.worldbank.org/indicator/AG.LND.PRCP.MM.

4. Given 0.7 meters of average annual rainfall, the mass of water falling on one square meter is 0.7 m × 1 m × 1 m = 0.7 m^3. Since the mass of water is 1,000 kg/m^3, the mass of the water is 700 kg. If that mass is deposited behind a hydroelectric dam

require 50,000 square meters of land area. A soccer field has an area of around 7,000 square meters, so the required land area is approximately the size of eight soccer fields.

Instead of considering just the area required to power one household, it is useful to consider the total electric power that could be provided if all of the rainfall in the United States were directed over 100-meter-tall dams. The land area of the United States is around 9,400,000 square kilometers, which is 9.4×10^{12} square meters. The average power per second that would be generated if all of the rain in the United States were collected behind a 100-meter-tall dam would be 2×10^{11} watts, while the average U.S. electricity consumption is around 5×10^{11} watts. Thus, even with the extremely optimistic premise that 100 percent of all the rain falling on the United States could be converted without loss into electrical power, hydroelectric power could not provide more than 40 percent of the electricity used by the United States or 7 percent of the total U.S. energy. A quick calculation based on the actual power production and catchment area of the Hoover Dam supports this estimate,[5] particularly if one includes the fact that some rainfall is absorbed by the earth or evaporates before it passes through the power plant.

Finally, in order to compare different renewable energy sources, it is useful to express their available energy in the same terms, so for all renewables I will calculate the available power per square meter. As discussed above, for hydroelectric power in the United States, the number is approximately 0.02 watts per square meter. A summary of quantities associated with hydroelectric power is provided in Table 8.

---

100 meters tall, then the energy resulting from that water falling over the dam is $mgh = 700 \text{ kg} \times 9.8 \text{ m/s}^2 \times 100 \text{ m} = 6,860$ joules/year ~ 0.02 W/m^2.

5. One way to test the accuracy of this estimate is to consider the Hoover Dam, which is about 200 meters tall, so it would produce 0.04 watts for each square meter of area that channels water into its dam. The catchment area of the Hoover Dam is approximately 400,000 square kilometers, which at 100 percent water exploitation would produce 16 gigawatts of power. In reality, it generates about 4 gigawatts, which means that 25 percent of all of the water raining into the catchment area generates electrical power by flowing through the dam.

4

ELECTRICITY FROM WIND

Hydroelectric power can not meet even the present U.S. electrical energy needs because the rainfall over the United States is insufficient, as a quick estimate in the previous section showed. However, on average, wind power could meet 100 percent of U.S. energy needs as will be discussed in detail in this chapter and as summarized by Table 9. In hydroelectric power, the gravitational potential energy in water behind a dam is converted to kinetic energy when the intake gate is opened and the water is allowed to flow. That kinetic energy is transferred to the rotor of a turbine, which makes the turbine spin. The spinning turbine is attached to a generator shaft, whose rotation produces electricity as a result of Faraday's law. As long as there is water behind the dam, electric power can be obtained at will by opening the intake gate. Opening and closing the intake gate allows dam operators to rapidly change the amount of power produced by the dam.

In wind power, the kinetic energy in the moving air molecules is transferred to the rotor of a turbine, which makes the generator shaft spin and produce electricity. Unlike hydropower, where operators control the water flow by opening or closing the intake gate, the air flow in wind power is almost entirely beyond our control. Not only is it beyond our control, but it is notoriously fickle: sometimes wind speeds are enormous, and sometimes the air is still.

In hydropower, the actual power provided by a hydroelectric plant was calculated by multiplying the maximum possible gravitational energy difference due to the flowing water times an efficiency factor. The efficiency factor is a measure of the gravitational potential energy that is not converted to electrical power. For hydroelectric dams, the efficiency factor is around 0.9.[1] Though it is possible to calculate the efficiency with which wind turbines convert wind power to electrical power, this is not the quantity that is usually emphasized in wind power calculations.[2] The most important feature of wind power is that the power generated depends on the wind speed, which we do not control; consequently, the power generated by a wind turbine varies strongly with time as a result of the changing wind speed. Thus, when calculating the power that a wind turbine can generate, it is convenient to average the generated power over all of the possible wind speeds. The capacity factor is a measure of the difference between the "nameplate" power of the turbine and the average power actually generated when the turbine is in a particular location.

The "nameplate power" is specified by the manufacturer. The nameplate power already includes the efficiency with which wind power is transformed into electrical power. The nameplate power represents the power generated when the wind speed is equal to the rated wind speed for the turbine, which is determined by the turbine design. The rated wind speed is typically around 10 meters per second. Of course, the actual wind speed is rarely equal to the rated wind speed. The capacity factor is defined as the ratio of the average wind power generated to the nameplate power. In other words, the average wind power = nameplate power × capacity factor, where capacity factors typically range from 15 to 40 percent.

1. "Reclamation, managing water in the west: Hydroelectric power," U.S. Department of the Interior Bureau of Land Management and Reclamation, July 2005, www.usbr.gov/power/edu/pamphlet.pdf.

2. The maximum efficiency of a wind turbine is called the Betz limit, and it is approximately 59 percent. It is discussed in detail in online Appendix 6 at http://thedata.harvard.edu/dvn/dv/HUP.

The detailed science governing wind power makes the electric power generation even more fickle than wind speed. I will show that the power generated depends on the cube of the wind speed. Thus, doubling wind speed increases the power by a factor of eight. As a result, a given particular turbine will generate much more power in a very windy location than in a less windy location. Even for a given wind turbine in a particular location at a given time, the power generated depends on the height of the turbine above the ground. Taking a particular turbine and placing it on a higher tower will increase the wind power since wind speed increases with height.

Unsurprisingly, the power generated by a wind turbine depends not only on the wind but on the properties of the turbine itself, the most important of which is that the power increases with the square of the rotor length. Thus, doubling the rotor length increases the power by a factor of four. If wind turbines are grouped together, one turbine can deflect wind from turbines behind it. Thus, the turbines in wind farms need to be spaced out, where the spacing is proportional to the rotor size.[3] As a result, though individual wind turbines can't be accurately characterized by power per square meter, wind farms can be.[4] Estimates based on existing wind farms suggest that the average power generated over area ranges from 2 to 10 watts per square meter for wind farms with

3. The Kentish flats wind farm has an area of approximately 10 square kilometers in which the wind turbines are spaced in a square grid with a separation of approximately 8 rotor diameters. The wind farm shows no reduction in wind power due to wind deflection. At 8–10 meters per second, the efficiency of the wind turbine is approximately 45 percent, which is about 76 percent of the theoretical Betz limit. See www.ref.org.uk. and www.wind-power-program.com/large_turbines.htm.

4. Caltech's John Dabiri reported findings that vertical axis wind turbines grouped together offer up to 30 watts per square meter of average power. The experiment used 10-meter-high vertical axis wind turbines with 1.2-meter diameters organized so neighboring turbines counter-rotate, where the turbines were separated by 1.65 turbine diameters. Potential order-of-magnitude enhancement of wind farm power density via counter-rotating vertical-axis wind turbine arrays, John O. Dabiria, *Journal of Renewable and Sustainable Energy* vol. 3, article 043104 (2011).

areas less than 10 square kilometers.[5] These wind power densities are consistent with estimates based on commercial 5-megawatt wind turbines;[6] however, recent theoretical work by David Keith and his colleagues suggests that very large 100-square-kilometer wind farms may yields as little as 1 watt per square meter as a result of the wind disturbances due to the presence of other turbines.[7] Optimizing the turbine type, size, and spacing for different sites and wind farm sizes remains a fundamental scientific challenge that has not yet been adequately addressed. There is even a question about what should be optimized. Possibilities include (1) average power over area; (2) residential electricity price required to make the investment viable, including any changes in transmission infrastructure; and (3) political/social acceptability of installing single wind turbines or wind turbine farms in different locations.

Wind power discussions are often very confusing because people use energy densities based on different assumptions. As the discussion above showed, reasonable assumptions result in wind power densities that can vary by a factor of 12,000, depending on whether one considers the 12,000 watts per square meter power over area of a single 5 megawatt wind turbine at 100 percent capacity factor[8] or the 1 to 10 watts per square

---

5. D. J. C. MacKay *Sustainable Energy—Without the Hot Air* (Cambridge: UIT Cambridge, 2009), p. 386.

6. For a 5 megawatt nameplate wind turbine with a hub diameter of 126 meters, assuming eight times diameter square spacing like that for the Kentish flats wind farm), gives a required surface area of $(126m*8)^2$. The power is the 5 MW/(surface area)*$0.3{\sim}1.5$ W/m^2, where 0.3 corresponds to a 30 percent average capacity factor, consistent with David Keith's work. In contrast, a three-rotor diameter spacing would give approximately 10 watts per square meter.

7. Are global wind power resource estimates overstated? Amanda S. Adams and David W. Keith, 2013 Environmental Research Letters vol. 8 article 015021. doi:10.1088/1748-9326/8/1/015021.

8. For a single wind turbine, the power/area is just the projection onto the ground of the volume swept by the wind turbine. For a 5 megwatt turbine with a 126-meter diameter and a 3-meter width, the projection on the ground is just $126\,m \times 3\,m = 378m^2$, so for a 30 percent efficient turbine, the average power delivered over area is approximately 4,000 watts per square meter. This is a hundred times the average power

meter average power generated by wind turbines in a farm where the turbines are spaced far enough apart that the wind from one does not block the other wind turbines. In addition, some places are windier than others, as shown by the world map at www.windfinder.com/weather -maps#2/-1.4/127.3. Such variations in wind speed may explain much of the large observed variation between the nameplate power of wind turbines and the average electric power actually generated by the turbines in wind farms. Some offshore wind farms have capacity factors of around 50 percent. For example, in 2012 the Alpha Ventus offshore wind farm showed a peak capacity factor of 50.8 percent,[9] and the Danish Horns Rev2 had a capacity factor of 52 percent.[10] In 2010 and 2011, Australian wind farms showed annual average capacity factors exceeding 30 percent,[11] which is consistent with the 30 percent factor used by the EIA and in this work. In 2014, Danish offshore wind farms show capacity factors above 40 percent with some exceeding 50 percent.[12]

Despite the large variation in wind power density, one can make conservative estimates that show that on average wind power could provide all of the energy requirements for the United States. Given that the complete energy needs of the United States require only 0.3 watts per square meter, if wind turbines generate an average of 1.5 watts per square meter, wind turbines on average could supply 100 percent of U.S. energy needs by covering less than 25 percent of the U.S. land area, and the U.S. electrical

---

delivered for a solar panel. Since this number scales linearly with the rotor length and the power scales as the square of the rotor length, this power density for a single wind turbine will increase linearly with the size of the turbine. In contrast, if the turbines must be spaced by a distance proportional to their rotor size, the power/area is independent of rotor size.

9. See www.lorc.dk/offshore-wind-farms-map/statistics/production/capacity-factor.

10. See http://energynumbers.info/capacity-factors-at-danish-offshore-wind-farms.

11. See http://windfarmperformance.info/documents/analysis/annual/2011/wind _capacity_factor_demand_2011.pdf.

12. See http://energynumbers.info/capacity-factors-at-danish-offshore-wind -farms, June 18, 2014. Raw data is available at www.ens.dk/node/2233/register -wind-turbines.

power demand could be met in less than 3 percent of the U.S. area. The National Renewable Energy Laboratory (NREL) has done a detailed calculation of the available wind power in the United States. In the calculation, researchers used actual measured wind speeds in different locations. They added up the available wind power for the entire United States, excluding areas that are unsuitable for wind power either because the wind power density is too low or the population density is too high. Their criteria excluded a bit more than 25 percent of the U.S. land area. This much more realistic calculation suggests that the United States could generate 11 terawatts of wind power[13] using the remaining 75 percent of the U.S. land area, whereas the total U.S. average power requirement is only 3 terawatts. Thus, the NREL estimates that providing power for the entire average consumption of the United States would involve approximately 20 percent of the U.S. land area, consistent with my cruder estimate.

Even with David Keith's very low 1 watt per square meter number, all U.S. electricity could be provided by around 5 percent of the U.S. land area, and all U.S. energy could be provided by 30 percent of the land area. It is important to note that the proposed turbine size and spacing turbine spacing are greater than those shown in Figure 4.1. One turbine per square mile is equivalent to eleven turbines in the area occupied by Central Park in New York City. If all of Manhattan were covered by wind turbines at that spacing, they would generate around 82 megawatts of power.[14] This would be grossly inadequate for just Manhattan's electrical power needs, which exceed 500 megawatts. A beautiful new study has mapped energy use in New York City zip code by zip code.[15] The largest

13. This assumes turbines on 80-meter towers. One hundred-meter towers could generate more than 12 terawatts.

14. The area of Manhattan is 59 square kilometers. One 5 megawatt wind turbine per square kilometer generates 1.4 megawatts of power, so the total power would be $59 \times 1.4 = 83$ megawatts of power.

15. http://modi.mech.columbia.edu/resources/nycenergy/. To convert kW-hours/m^2/year to W/m^2, multiply by 0.114. In the *Huffington Post* version, the popup is not

power consumption on the map corresponds to around 570 watts per square mile, which is more than 400 times the power that could be provided by wind turbines despite people in New York City consuming substantially less electricity per capita than the national average.[16]

Considering renewable energy in the context of Manhattan highlights the issue of population density, which plays a crucial role in determining whether a particular region can meet its energy needs using renewable resources located within its borders. For example, the U.K. population density of 650 people per square mile poses greater challenges for renewable energy substitution than the eighty-four people per square mile in the United States. In the United States, wind power can easily provide complete average energy substitution. In the United Kingdom, wind power could barely provide the average energy if the country were completely covered in wind turbines, even though the United Kingdom uses about half the energy per capita of the United States. If U.K. residents consumed as much energy per capita as residents of the United States, wind power could not possibly provide enough energy. Of course, wind power can be augmented by solar power and other renewable energy sources; however, the fundamental issue of the relationship between energy and population density remains.

When attempting to determine whether it is feasible or desirable to convert to wind power, it is extremely important that the land area below wind turbines can still be used for other purposes, such as livestock grazing or crop growth, as illustrated in the photograph in Figure 4.1. For a 5 megawatt wind turbine farm with a spacing of three times the rotor diameter, the area occupied by the tower is less than 0.1 percent of

---

blocked by the legend: www.huffingtonpost.com/2012/02/02/new-york-city-energy-interactive-map_n_1249856.html.

16. New York City residents also use substantially less total energy per capita and emit less carbon per capita than the U.S. national average. See www.hks.harvard.edu/var/ezp_site/storage/fckeditor/file/pdfs/centers-programs/centers/taubman/policybriefs/greencities_final.pdf.

Figure 4.1   Cows grazing in a field of horizontal-axis wind turbines on a good solar power day. The picture was taken in Germany by Dirk Ingo Franke. An important difference between wind and solar power is that with the exception of the land directly under the bases of the turbine towers, all of the land is still available for agriculture, including raising crops and grazing animals. In contrast, the land below solar panels is in complete darkness, so crops cannot grow and animals cannot graze. This is an important distinction when considering how much land area is "required" to provide energy using wind power versus solar power.

---

the area occupied by the wind farm. When service roads and other land use requirements are included, around 3 percent of the land area of a large wind farm is taken out of agricultural use.[17] This feature was ignored in the recent political drama that occurred as a result of someone in the

17. See http://css.snre.umich.edu/css_doc/CSS07–09.pdf.

U.K. government posting a graphic indicating the vast area that wind turbines or solar panels must cover in order to produce as much average power as the Hinkley Point C nuclear power plant.[18] The graphic was quickly removed from the government site, but it provoked a substantial debate about how to fairly compare renewable energy with nuclear energy or even conventional fossil fuel–generated energy.

In sum, even using the most conservative wind power estimate, existing technology would easily allow the average output of wind generators to supply 100 percent of the average energy needs of the United States in an all-electrical energy economy, where all climate control and transport, with the exception of aviation and some industrial processes, are powered by electricity. Such a system is completely sustainable, emits negligible pollution, and requires absolutely no energy imports. Assuming that concerns involving aesthetics and noise pollution are resolvable, wind power would seem to be the answer to all of the world's energy concerns, if cost were not an issue. Current wind turbine pricing is around $1–$2 per nameplate watt for large wind turbines; consequently, at 30 percent capacity factor, providing all U.S. energy would require an investment of about 50 percent of the U.S. GDP, whereas providing all of the electrical energy would require around 7 percent. These are daunting numbers but not outside the realm of possibility.

The paragraph above considered capital costs for the turbines themselves, but the desirability of shifting to a wind-powered energy system also depends on the overall cost of the system. The overall cost includes not only the cost of purchasing the turbines but also the maintenance costs, land use costs, and delivery costs required to transport electricity from wind farms to consumers. Just such a calculation has been made by the U.S. Energy Information Agency, and the results are shown in Figure 4.2. The graph shows that the electricity prices from wind farms are somewhat better than those from coal-burning electrical power plants,

---

18. See www.ibtimes.com/hidden-cost-solar-wind-power-one-image-uk-1441400.

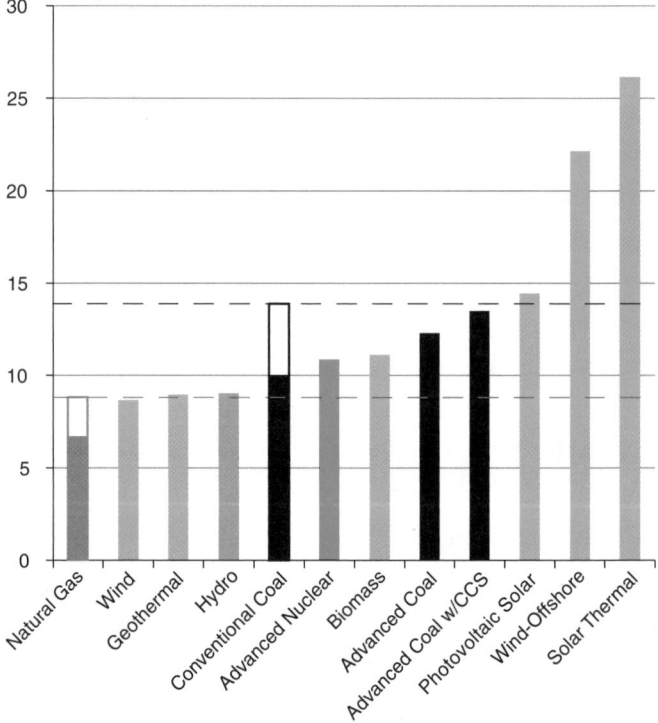

Figure 4.2 2013 U.S. Energy Information Administration estimates of the cost per kilowatt-hour of new electricity-generating facilities that will enter service in 2018. The solid bars show the cost without including carbon. Three bars are shown for coal, indicating the cost increases associated with the strategies designed to reduce the emissions associated with burning coal. Advanced coal includes various mitigation strategies, but it does not include carbon capture and storage (CCS). Information on some strategies for mitigation is available at www.worldcoal.org/coal-the-environment/coal-use-the-environment/improving -efficiencies/. Without carbon, natural gas has the lowest cost. The open rectangles show the costs with carbon for natural gas and conventional coal with the carbon cost included. For comparison with other costs, the horizontal pink and black dashed lines indicate the costs of natural gas and coal, respectively, when carbon is included. It is interesting that the cost of conventional coal with carbon included is approximately equal to the actual cost of advanced coal with carbon removal included. With carbon included, wind power has a slightly lower cost than natural gas, and conventional coal is so expensive that only offshore wind and solar thermal energy show significantly higher prices. Estimated levelized cost of new generation resources, 2018 available at www.eia.gov/forecasts/aeo/er/electricity_generation.cfm, document number DOE/EIA-0383ER, table 1.

which suggests that switching from coal to wind is advantageous on merely economic grounds without factoring in the substantial deleterious effects associated with coal-fired plants. Furthermore, land-based wind power is more cost effective than any power source except burning natural gas, though hydroelectric and geothermal power have almost the same cost as land-based wind. Thus, anyone choosing to build a new U.S. electrical power plant based only on cost would build a gas-fired plant. Building a wind power plant would become economically advantageous only if additional costs, such as government subsidies and carbon emissions, are factored into electricity prices.

Debates continue about how to calculate the total government subsidies provided to different types of power. A broad definition of subsidies must include tax advantages for drilling or mining fossil fuels, as well as guarantees provided for nuclear plant owners. An expansive definition of government subsidies would include geopolitical costs of our relationships with fossil fuel–exporting countries. In addition to government subsidies, cost must include the social penalties associated with energy extraction and transport, as well as the costs of all the emissions that result from consuming energy. Similarly, the cost of wind power would include government subsidies for the development of long-distance HVDC lines and "smart grid technologies."

Again, such a calculation is well beyond the scope of this book, but I will provide a simple calculation that adds an estimated carbon dioxide emission cost to the prices shown in Figure 4.2. The Environmental Protection Agency (EPA) estimates that burning natural gas generates 117,000 pounds of carbon dioxide per billion British thermal units, whereas oil generates 164,000 pounds, and coal generates 208,000 pounds.[19] Given the U.S. government estimate of a social cost of $37 per ton and a 33 percent energy efficiency in conversion of fuel to electricity, one can calculate the additional cost associated with carbon emission. The results are

19. See www.epa.gov/cleanenergy/energy-resources/refs.html).

2.2, 3.1, and 3.9 cents per kilowatt-hour for natural gas, oil, and coal, respectively. Figure 4.2 shows the levelized cost for different energy sources when the $37 per ton social cost is included. Adding costs for other pollutants, such as particulates, and sulfur and nitrogen oxides would increase the costs of electricity generated by fossil fuel burning, creating the worst emissions in all categories, as shown in Figure 6.3. Adding costs associated with energy extraction and indirect government subsidies for fossil would further favor renewable energy; however, if subsidies for fossil fuels were included, subsidies for renewables would have to be included as well.

Independent of cost there is an extraordinarily serious fundamental scientific issue with wind power: when the wind is not blowing, no wind power is generated and even when the wind is blowing the power generated is a highly nonlinear function of the wind speed. This is the crucial problem with wind power. Wind speed varies on several time scales. The English word "gust" describes a variation in wind speed that lasts for only a very short time, and daily wind speed variations plague sailors and wind surfers. Finally, seasonal wind variation is one reason that kite flying in the spring is more popular than kite flying in the middle of the summer, despite the better weather in the summer. For example, at Logan Airport in Boston, the average wind speed in March and April is 14 miles per hour, whereas the average speed in June and July is only 10 miles per hour, which means that in the spring a wind turbine at Logan would generate almost three times more power than it would generate in the summer.[20]

These problems with changing wind speed could be overcome if we had sufficient energy storage to contain the excess energy generated when the supply of wind power exceeded demand, though most people think more of storage lasting around a day, rather than storage lasting over a season. Another alternative is to have a sufficient number of wind turbines distributed over a wide enough geographical area that the power generated is always sufficient. This would require a much larger number

20. See www.windfinder.com/windstats/windstatistic_boston_logan_airport.htm.

of wind turbines than the estimate above indicated, and it would require a revolution in the electricity distribution system, possibly including a broad international network that connected wind generated in one country to consumers in other countries. Both alternatives are less practical than the systems that are employed now in which the electrical generating capacity of wind power must be backed up by a fossil fuel–burning electrical generating system. Such duplicate generating capacity ties up a great deal of capital, resulting in hidden additional costs associated with conversion to renewables. In addition, the fossil fuel backup system suffers from all of the issues associated with fossil fuel burning, though all of the negative impacts are reduced during the times when the electricity is provided by the wind. Backing up wind power with hydroelectric power overcomes the emission issues associated with burning, but as discussed in Chapter 3, the amount of rainfall is too small for hydroelectric power to provide much more than the 3 percent that it currently contributes to total U.S. energy consumption. Even 3 percent of the total power corresponds to 30 percent of the total power 10 percent of the time, which is important in considering hydroelectric power as a backup for wind. In addition, because of all of the energy losses associated with burning, renewables only need to supply around 50 percent of the total U.S. energy consumption, which would allow hydro to provide 10 percent of total U.S. energy 6 percent of the time.

An alternative to providing a backup for wind power during low wind times is to provide excess wind power during average generation times, creating periods where energy supply significantly exceeds demand. Fortunately, energy uses that do not require constant power, such as desalination, hybrid car battery charging, or generation of hydrogen-using hydrolysis might provide excellent opportunities to profit from periods during which the wind power production exceeds the demand, which would reduce the cost of maintaining "excess" wind capacity to provide sufficient power during low wind production periods. Below I discuss why wind power is so much more variable than any other power source and why the siting of wind power plants is so much more important than the siting of any other power plant.

OVERVIEW OF WIND POWER

Like hydroelectric power, wind power has its origin in sunlight that is delivered to the earth. Wind blows because sunlight warms some areas of the earth more than others, resulting in a transfer of air between regions at different temperatures. Thus, in the case of wind power, sun light shining over large areas provides energy that accumulates over time and space. In hydroelectric systems, the damming of flowing water provides natural energy storage. The system stores energy as gravitational potential energy that is controlled by the gate that determines how much water flows out of the dam. As a result, though rainfall is intermittent, including dry periods that can last for months, the water behind dams can collect over such long times that fluctuations in rainfall do not result in fluctuations in hydroelectric power; therefore, hydroelectric power can provide "constant" electrical power, even though the sunlight that ultimately powers the system has large variations. To date, such hydroelectric systems represent the simplest and most efficient method of storing enormous amounts of energy. Similarly, one of the reasons that fossil fuels are so widely used is that they too have stored solar power over long periods of time and they can release that stored power on demand when they are burned while providing fairly safe and inexpensive energy storage when no power is required.

Such natural energy storage features are not present in wind power systems or photovoltaic systems: if the electricity generated is not immediately used, it is lost. The absence of effective electrical storage poses significant challenges in wind- and solar-powered systems. As a result, even if such systems can provide more than 100 percent of the average power required by the United States, they cannot be used as the exclusive power sources because there are times when wind and solar provide little or no power. At those times, electricity must be generated using stored power.[21] The storage can be reusable, such as dammed water, or it can be nonreusable, such as the chemical energy stored in gasoline.

Wind power is a very old power source, long predating electrical power generation. Wind power was originally used to lift water up from wells and to grind grain. In the twentieth century, wind power was used to generate electrical power in some isolated areas, but it has not been a large-scale source of electrical power because hydrocarbon burning has provided relatively inexpensive, easily controlled power. In the last thirty years, wind power has enjoyed a renaissance because it can provide electrical power without generating carbon dioxide or nuclear waste, so countries concerned with the environment encourage wind power. Furthermore, wind power is available within the national boundaries of countries that lack fossil fuel resources, so wind power allows such countries to reduce energy imports.

The discussion above presented calculations indicating that it is technically possible for wind power located with the borders of the United States to provide 100 percent of the total average power consumed in the country. Of course, providing 100 percent of the electrical power is significantly less demanding. The theoretical calculations can include all sorts of mistaken assumptions; therefore, it is important to consider cases where wind power provides a substantial portion of national electricity consumption in major industrialized nations. At present, wind already provides the nation of Denmark with more than 25 percent of its electrical power. More dramatically, on November 3, 2013, the electric power generated by wind turbines exceeded the national electricity consumption.[22] In other words, wind power provided more than 100 percent of

21. In the case of solar power, there will always be times when the continental United States will be in darkness; however, pooling wind power over the entire United States might virtually eliminate nonseasonal wind power fluctuations even without storage.

22. See http://energytransition.de/2013/11/denmark-surpasses-100-percent-wind-power.

the total electricity requirement, supporting the idea that Denmark could achieve its commitment to eliminating fossil fuel consumption entirely by 2050.[23] Even within the United States, Iowa and South Dakota generate more than 20 percent of their electricity using wind power. As shown in Tables 5 and 6, U.S. total energy use is around seven times its electricity use; therefore, the calculations probably do correctly predict that the United States could comfortably generate 100 percent of its average total energy consumption using wind power. In the absence of energy storage, meeting average power requirements is very different from meeting peak power requirements; therefore, in the rest of this chapter, I will discuss fundamental features of wind power arising from basic science that make complete fossil fuel replacement nontrivial.

BASIC SCIENCE

Most people have fond memories of the magic of toy pinwheels spinning as the wind blows past them or as an excited child waves them while running through still air. Those pinwheels follow the same basic physics as wind turbines. This section will discuss the fundamentals of extracting electrical energy from the wind and in particular the location of wind turbines, which is so important, and the reason why the length of wind-turbine blades has been increasing as wind-power technology has improved.

Wind turbines are divided into two broad categories: vertical axis wind turbines and horizontal axis wind turbines. In vertical axis turbines, the axis of rotation is perpendicular to the surface of the earth, and in horizontal axis turbines, the rotation axis is parallel to the surface of the earth. In other words, in a vertical axis wind turbine, the rotor moves in a plane parallel to the surface of the earth, and in a horizontal wind turbine, the blades move in a plane perpendicular to the earth's surface, as illustrated in Figure 4.3. Photographs of actual vertical and horizontal axis wind turbines are also shown in Figure 4.3.

23. See http://denmark.dk/en/green-living/wind-energy/.

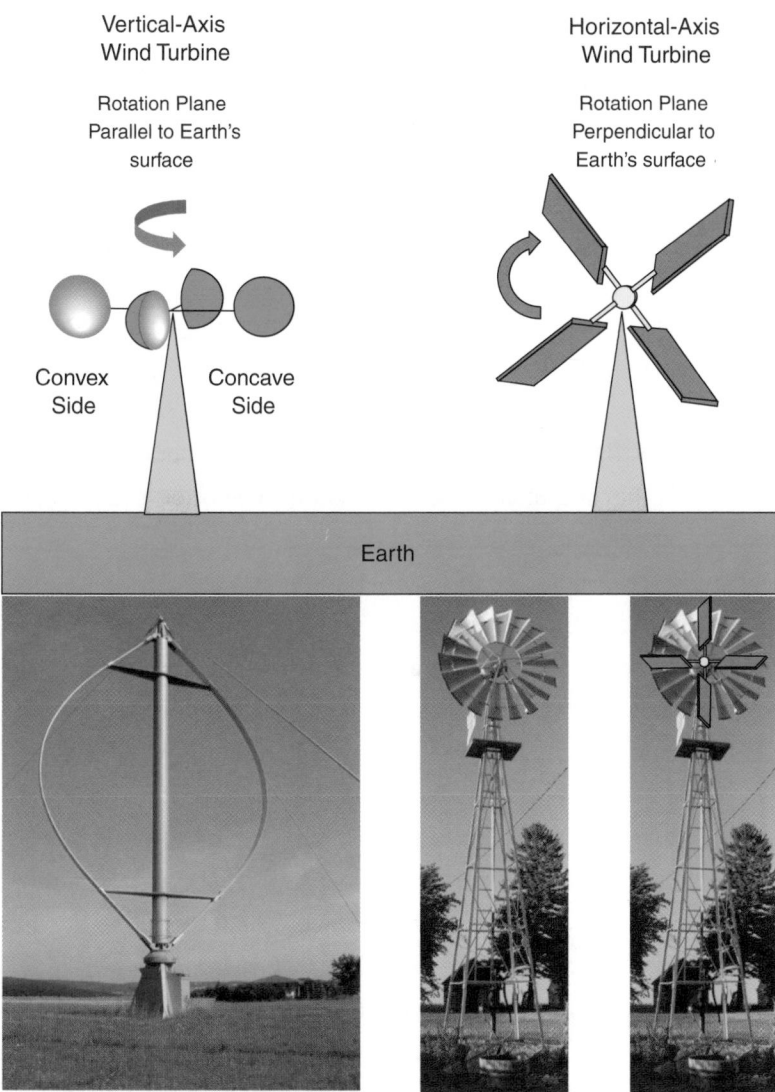

Figure 4.3 Vertical- and horizontal-axis wind turbines. The vertical-axis wind turbine schematic is a classical anemometer. Modern vertical-axis wind turbines often resemble egg beaters, as shown in the lower left image. The image at the lower left is a very large Darrieus wind generator near Heroldstatt, Germany, photographed by W. Wacker. The schematic on the upper right resembles a classic Dutch windmill. The images in the lower right are photographs from Ben Franske, with the schematic windmill blades superimposed to illustrate orientation.

Vertical Wind Turbines

Figure 4.3 shows a very simple wind power system consisting of four half spheres. Most people would refer to the system as an anemometer rather than a wind turbine. As an anemometer, it is used to measure wind speed rather than to harvest electrical power; however, it provides an opportunity to explain how wind can make a shaft rotate.

Anemometers have the wonderful feature of working regardless of wind direction, as long as the wind direction is parallel to the surface of the earth. To understand how they work, consider people playing tug of war. If they tug equally, they are both working very hard, but no one is going anywhere. If they tug unequally, there will be a net motion in the direction of the greater tug, as illustrated in Figure 4.4.

A similar result holds, if, instead of pulling on a rope, people compete by pushing on a plane propeller, as shown at right in Figure 4.4. If they push equally hard, they are doing a lot of work, but the propeller won't move. If only one person is pushing, the propeller will spin.

In the wind system, the wind hits the half spheres and pushes on them. If the wind pushed just as hard on the concave side of the sphere as on the convex side, then the force exerted by the wind would not make the shaft rotate because it was pushing equally hard on both sides. In contrast, if one side pushes harder, there will be net motion in the direction of the harder push, as illustrated in Figure 4.5. In the case of the half spheres, the wind pushes harder on the hollow (concave) side than on the convex side because wind hitting the concave side is sent back in the original wind direction, whereas wind hitting the convex side bends around the cup but keeps going in the initial direction; consequently, the shaft rotates in the direction that pushes the hollow side forward, regardless of the wind direction. This effect is illustrated in Figure 4.5, which shows that the wind cup rotates regardless of the wind direction. A much more detailed discussion of vertical axis wind turbines is presented in online Appendix 6 at http://thedata.harvard.edu /dvn/dv/HUP.

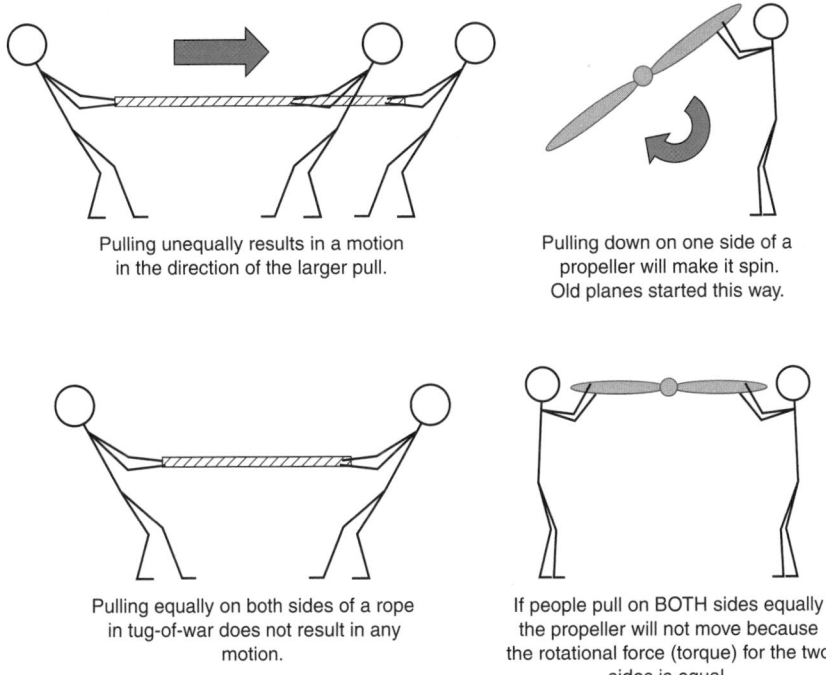

Pulling unequally results in a motion in the direction of the larger pull.

Pulling down on one side of a propeller will make it spin. Old planes started this way.

Pulling equally on both sides of a rope in tug-of-war does not result in any motion.

If people pull on BOTH sides equally the propeller will not move because the rotational force (torque) for the two sides is equal.

Figure 4.4 The left column illustrates how no net motion will result if an object is being pulled equally from both sides, but motion will occur if one side is pulled harder than the other, as indicated by the two people pulling from one side. This is the principle behind tug-of-war games. The right column shows the analogous effect for a rotating system. If the torque on both sides is the same, there will be no rotation; however, if the torque on one side is larger, the propeller will rotate.

Though anemometers can generate electrical power, they are not efficient converters of wind power to electrical power because the cups with their convex sides facing the wind still experience a significant wind push that offsets the amount of power delivered to the cups whose concave side faces the wind. Vertical wind turbines are still sometimes used in small wind generators where expense and simplicity are more important than overall energy efficiency.

Current vertical axis wind turbines are much more elegant. They do not use cups, buckets, or troughs to redirect the wind. Instead, they use

Figure 4.5 Illustration of why anemometers turn regardless of the wind direction. The left shows the effect of torque when the wind is blowing from the left. The concave cup feels approximately twice the force of the convex cup, so the anemometer rotates in the direction where the concave cup is pushed back by the wind. The illustration on the right shows what happens when the wind is blowing in the opposite direction. Now the lower cup is concave in the direction of the wind, so the rotation direction is opposite to the direction shown at left; however, the system still rotates.

blades, similar to the propeller blades on planes.[24] Recent work has suggested that vertical axis wind farms can have power densities on the order of or in excess of solar panels, making vertical wind turbines suit-

24. NASA offers a program called foilsim, which provides a wonderful simulation of the motion of air over wings or blades.

able for urban and near urban areas; however, in all vertical axis wind turbines, the wind pushing on one side of the turbine always opposes the dominant motion on the other side. This lowers the efficiency with which wind power is converted to electrical power. In contrast, when wind passes by a horizontal wind turbine, every blade is pushed in the same direction, allowing horizontal wind turbines to be more efficient than vertical wind turbines, which is one reason that large wind farms almost always use horizontal turbines.

Horizontal Wind Turbines

Unlike vertical wind turbines, where the rotor axis is perpendicular to the surface of the earth, in horizontal axis turbines, the axis around which the blades rotate is parallel to the surface of the earth. When most people picture wind turbines, they picture horizontal axis wind turbines, such as the traditional four-blade Dutch windmills or multiblade windmills from the American West, which consist of flat boards attached to a central shaft.

Modern-day horizontal axis designs typically have two or three curved blades attached to two sides of a shaft. For the purpose of simple illustration, I will first consider a four-blade system, illustrated in Figure 4.3 alongside a photograph of a Western windmill to connect the cartoon with a real system.

Figure 4.6 shows a diagram of the four-blade system. It illustrates how air that blows parallel to the surface of the earth can make a windmill rotate. In the image, the wind is moving perpendicular to the plane swept out by the face of the windmill blades. The plane swept by the blades is parallel to the surface of the printed page, so the wind can be described as blowing toward or into the surface of the page. When the air molecules in the wind hit the blade, they are redirected. The black arrows indicate the direction of the air molecules in the wind after they have hit the turbine blades. The black arrows lie in the plane of the page, thus air that had been moving in the direction perpendicular to the page is redirected

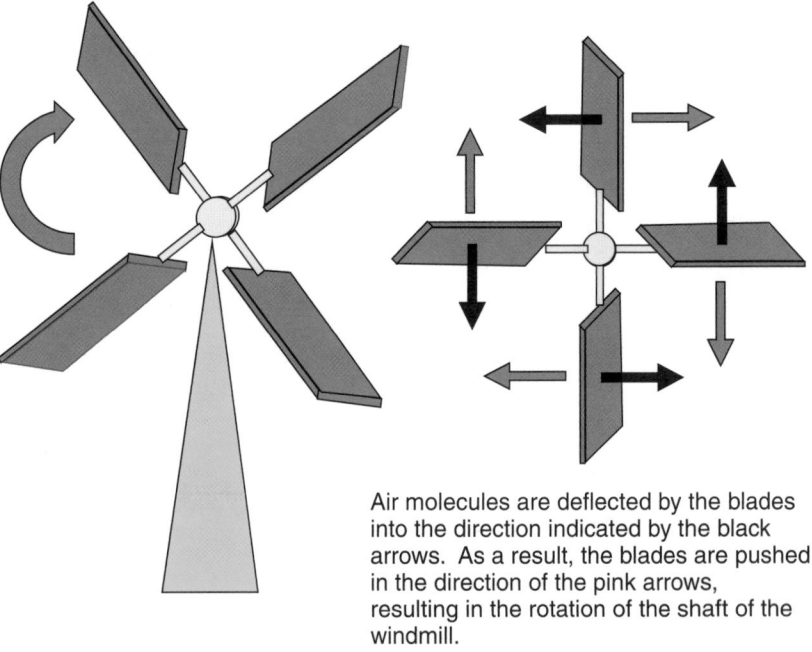

Wind blowing into the page

Air molecules are deflected by the blades into the direction indicated by the black arrows. As a result, the blades are pushed in the direction of the pink arrows, resulting in the rotation of the shaft of the windmill.

Figure 4.6 Air molecules moving toward the blades as a result of wind are redirected when the wind is parallel to the rotation shaft, which is the direction that is perpendicular to the printed page. The black arrows indicate the direction of air molecules in the wind after they have collided with the blades. Collisions with the blades redirect the air molecules so that their motion has a component in the plane of the page. From conservation of momentum (online Appendix 3.4.2 at http://thedata.harvard.edu/dvn/dv/HUP). As a result of momentum conservation, the redirection of that wind by the blades causes the shaft to rotate, just as the redirection of the wind by an anemometer causes the shaft to rotate; however, in the anemometer, the interaction between the convex cups and the wind resisted the motion induced by the wind pushing on the concave cups, whereas in the horizontal wind turbine, as in a child's pinwheel, the wind induces the same rotation direction in all of the blades.

into the plane of the page. Conservation of momentum dictates that if the blades push on the air molecules and change their direction, then the blades will be pushed in the direction opposite the change in direction.[25] The magenta arrows indicate the push that the blades receive in the direction parallel to the plane of the page. The magenta arrows indicate that the redirection of the wind exerts a force that makes the blades rotate. Given the 45-degree angle, the blades not only experience a force that makes them rotate, they are also pushed into the page.

This push that results from the redirection of the wind can be seen by placing a notecard or piece of cardboard in front of your face with the card in the orientation illustrated in Figure 4.7. Blowing on the card will push it up and back because the card pushes the air back and down. If you blow and release the card, it will fly up and away from you. This effect allows kites and planes to fly. In the case of a wind turbine, the component of the force associated with the card flying forward, away from your face, simply pushes on the wind turbine. It does not contribute to the power; however, the component of the force associated with the upward motion of the card is the component that makes the turbine rotate. As indicated in Figure 4.6, this component of the force induces the same rotation direction in every blade of the windmill.

Thus, unlike the cup system, where the wind pushed half of the cups in a direction that opposes shaft rotation, the rotation direction induced by the wind is the same for all of the blades in the turbine. As a result, horizontal axis wind turbines can be more efficient than vertical axis turbines; however, vertical axis wind turbines work regardless of the wind direction. In contrast, as shown in Figure 4.8, horizontal axis systems work only if the wind is blowing parallel to the shaft.

Users of horizontal axis windmills can respond to this problem in various ways. The most common response is to allow the direction of the shaft to rotate so it is always parallel to the wind. In classical windmills

---

25. Conservation of momentum is discussed in detail in online Appendix 2.

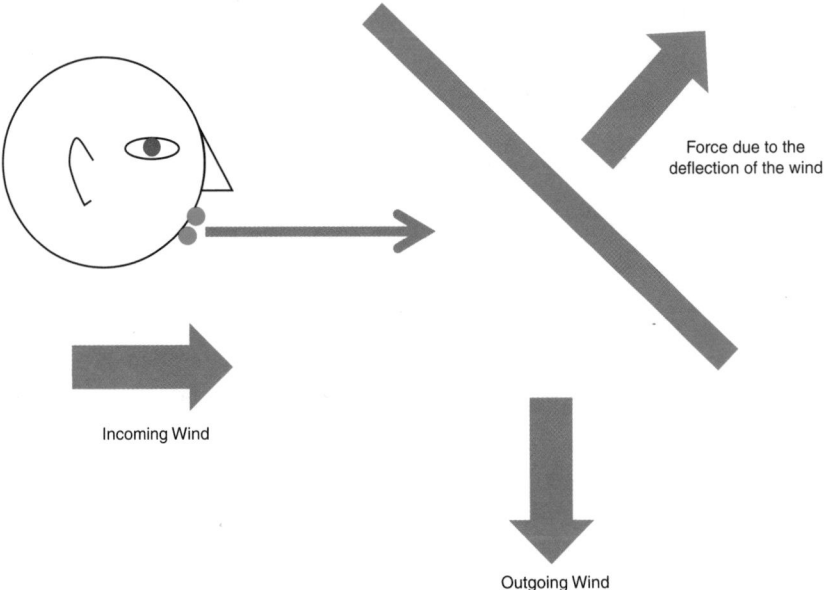

Force due to the deflection of the wind

Incoming Wind

Outgoing Wind

Figure 4.7 Illustration of the deflection of a stiff piece of paper by a person blowing air toward the paper. Initially the air moves perpendicular to the person's face. After the moving air collides with the paper, it moves down. As a result of conservation of momentum, the decreases in the air velocity in the forward direction push the piece of paper back away from the person. Similarly, the increase in the velocity in the direction of the person's feet, pushes the piece of paper up. This same effect makes kites fly up and away from the person holding the kite string as the kite redirects the wind velocity away from the initial wind direction and toward the ground. The angle between the kite and the wind determines the ratio of the force that lifts the kite up to the force that pushes the kite in the original direction of the wind.

from the American West, this is achieved by a tail attached to the back of the windmill.[26]

The efficiency of a wind turbine depends not only on the angle between the rotating shaft and the wind but also on the blade angle. In Figure 4.9, a 45-degree angle had been chosen. If the direction of the long side of

26. A second option is to fix the wind turbine direction but locate the turbine in a place where the wind blows predominantly along one direction—for example, near the seashore, where the wind predominately blows perpendicular to the coastline.

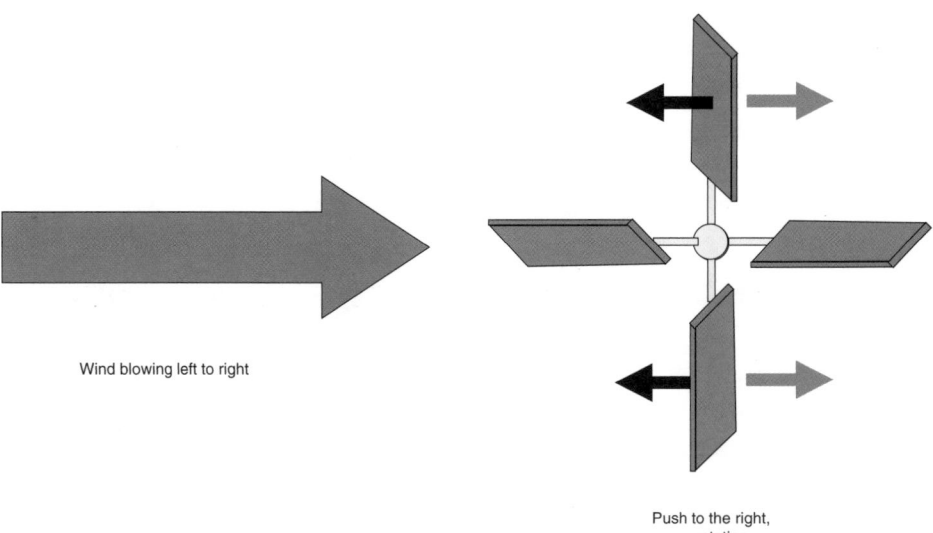

Wind blowing left to right

Push to the right,
no rotation

Figure 4.8 Horizontal-blade windmills do not rotate if the wind is blowing perpendicular to the shaft. In this diagram, the wind is initially moving in the plane of the page in the direction indicated by the large blue arrow. Again, the black arrows indicate the direction of the wind after the air molecules collide with the blades. The wind simply blows past the blades at the 3 o'clock and 9 o'clock positions, so those molecules do not contribute to rotation. For the blades at 12 o'clock and 6 o'clock, the change in wind direction within the plane of the page pushes both blades in the wind direction. Thus, the blade a 12 o'clock would produce clockwise rotation, whereas the blade at 6 o'clock would produce counterclockwise rotation. The two pushes are equal and opposite. This is the situation depicted in the bottom right illustration in Figure 4.4, which shows that there is no net rotation.

the blade is perpendicular to the shaft direction, then the force on the blades is large; however, the force simply pushes the windmill backward. The blades would not rotate at all. Similarly, if the direction of the long side of the blade is parallel to the direction of the wind, then the blades will still not rotate, but the force exerted on the windmill will be much smaller because the blade area facing the wind is smaller. A similar reduction in force occurs when carrying a flat sheet of cardboard on a windy day. Walking is much easier if the long direction of the cardboard is parallel to the direction of the wind.

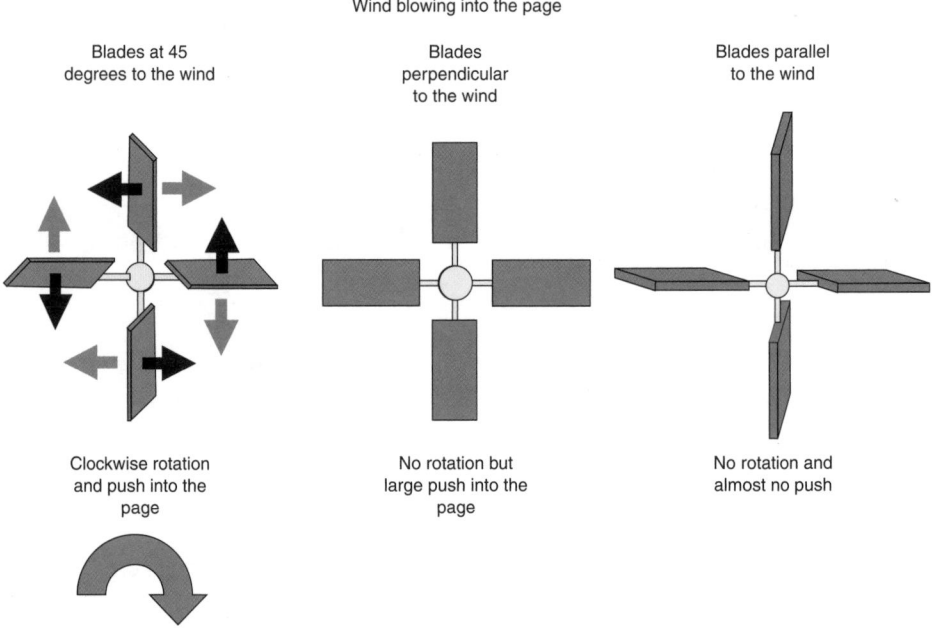

Figure 4.9 Illustration of the effect of blade angle on how effectively wind turbines with flat blades can translate wind energy into rotational energy. If the blade angle is approximately 45 degrees, as shown in the left-most image, then some of the wind force will be wasted pushing against the tower; however, the rest of the wind force will result in a rotational motion of the blades, as shown in Figure 4.6. In contrast, the central figure shows that if the blades are perpendicular to the wind direction, then the wind will provide the largest possible force on the blades, but the blades will not turn; they will simply push the tower back. Finally, the right most figure shows that, if the blades are parallel to the direction of the wind, the wind exerts almost no force. Real turbines have curved blades, like airplane propellers. The net force on such curved blades is much more difficult to calculate; however, the same basic ideas underlie those more complex calculations.

The discussion above considered only flat blades, but more effective windmills use curved blades that increase the transfer of energy. The curved blades are similar to plane propellers.[27] Much effort has gone into

27. In the case of a plane propeller, the air is stationary and mechanical work is done to spin the propeller blades. The spinning propeller blades make the air move from the front of the plane to the back of the plane. Thus, the air molecules acquire a

optimizing the shape of wind turbine blades, and existing wind turbines already approach the theoretical efficiency limit; therefore, blade design is not believed to represent an area where significant improvements can be made.

Scaling of Wind Power with the Square of the Rotor Size

As discussed above, wind turbines turn because they redirect the wind. From conservation of momentum if a rotor pushes the air to the right, then the air will push the rotor to the left. The left half of Figure 4.10 shows a wind turbine whose blades have a radius R, while the illustration on the right shows that four wind turbines with radius R/2 can fit in the same area as the single wind turbine with radius R. If every air molecule passing through the turbine hits one of the blades, then four times as many molecules hit the turbine with radius R as hit one turbine with radius R/2.

Figure 4.10 is inexact, but the rule that the wind power from a turbine scales as the square of the length of the blade is so important that it is worth deriving more rigorously. If the molecules are considered to be independent, a molecule can transfer momentum to the rotor only if it hits the rotor. Thus, the molecule must past through the circle that the rotor sweeps out as it turns around. If the rotor has a radius R_{rotor}, the circle has a radius πR_{rotor}^2. Thus, the wind power scales as the square of the radius. This is the effect that has led manufacturers to continue to increase the radius of the wind turbines: doubling the rotor length produces four times the power, and tripling the rotor length produces nine times the power. As long as the cost of the wind turbine increases by a smaller factor, increasing wind turbine size will lower the cost of power generation. For wind turbines up to 5 megawatts, this rule seems to hold.

---

velocity that pushes them backward along the plane. Conservation of momentum implies that the plane will then acquire a forward motion because the propeller blades push the originally stationary air back. Of course, a stationary propeller does not exert any pull on a plane, just as wind turbine rotors do not rotate if the wind is not blowing.

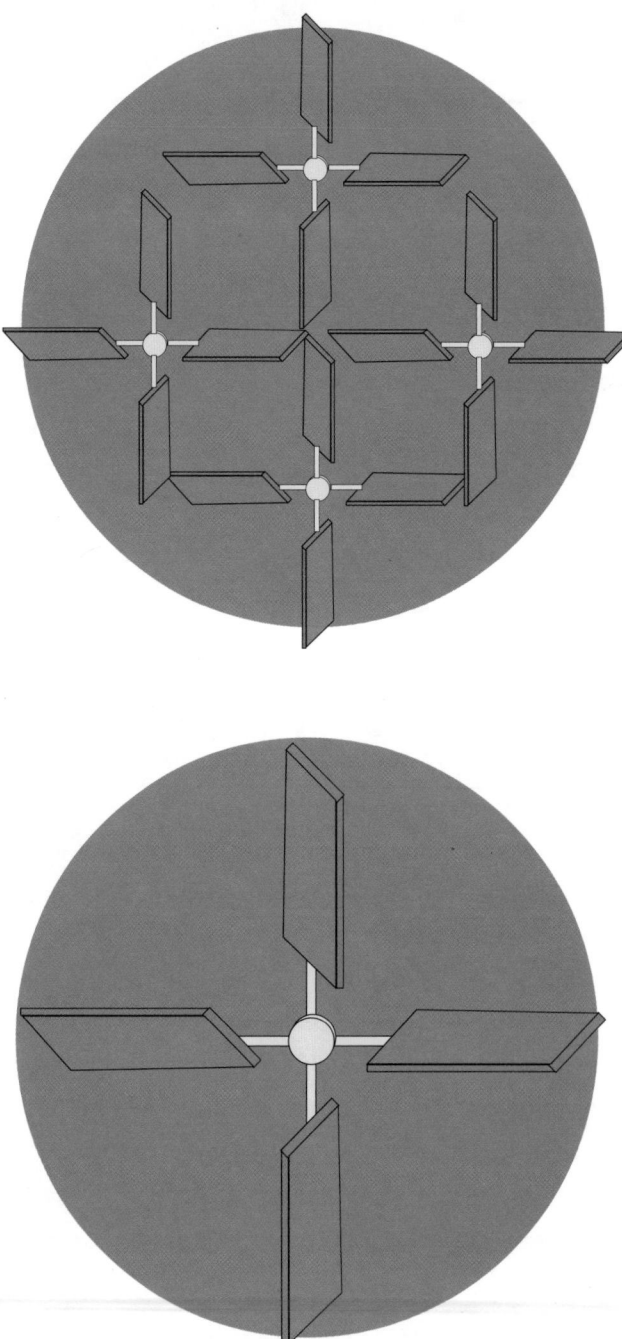

Figure 4.10 A simple illustration of why the total number of air molecules that collide with a wind turbine increases to the square of the length of the turbine blades. This figure compares the area swept by one turbine with a blade length R, to the area swept by four turbines, each of which has a blade length of R/2. The area through which one wind turbine with radius R rotates is the same as the area over which four turbines with radius R/2 rotate.

Scaling of the Wind Power with the Cube of the Wind Speed

Each air molecule with a velocity v can transfer $1/2 \, m \times v^2$ of energy to the rotor, as discussed in online Appendix on 3.1. Thus, one can calculate the maximum power that could be delivered by the wind by multiplying the maximum energy per molecule, $1/2 \times m \times v^2$, by the number of molecules hitting the turbine in one second. The maximum number of molecules that could hit the turbine during a time T is given by the total number of molecules passing through the $\pi \times R^2_{rotor}$ area swept by the turbine during a time T. This number, Number (T), is equal to the air density times the volume occupied by the air passing through the swept area during the time T.

If all of the air molecules moved at the wind velocity v_{wind},[28] this volume would be given by the volume of the cylinder with a radius equal to the R_{rotor} and a length given by $v_{wind} \, T$, which is the product of the wind velocity v_{wind} and the interaction time, T, as shown in Figure 4.11. Thus, the total number of molecules that can interact with the blade during a time T is given by

$$\text{Number (T)} = \text{density} \times \pi \times R^2_{rotor} \times v_{wind} \times T,$$

as illustrated in Figure 4.11. Thus, the number hitting the blade during a time T is proportional to the density of the air molecules times the square of the rotor radius times the wind speed times the time.

Given that each air molecule moving at velocity v_{wind} can deliver a maximum energy of $1/2 \, m \times v^2_{wind}$, the maximum energy that could be delivered in a time T is $1/2 \, m \times (v^2_{wind})$ Number (T) = density $\times \pi \times R^2_{rotor} \times v^3_{wind} \times T$. Power is defined as energy per unit time, so the power contributed is given by the energy contributed during time T, divided by T, which is equal to density $\times \pi \, R^2_{rotor} \times v^3_{wind}$.

28. Air molecules actually have very large random velocities that make no contribution to the overall motion of the rotor. The only net contribution comes from the net motion, which is given by the wind velocity.

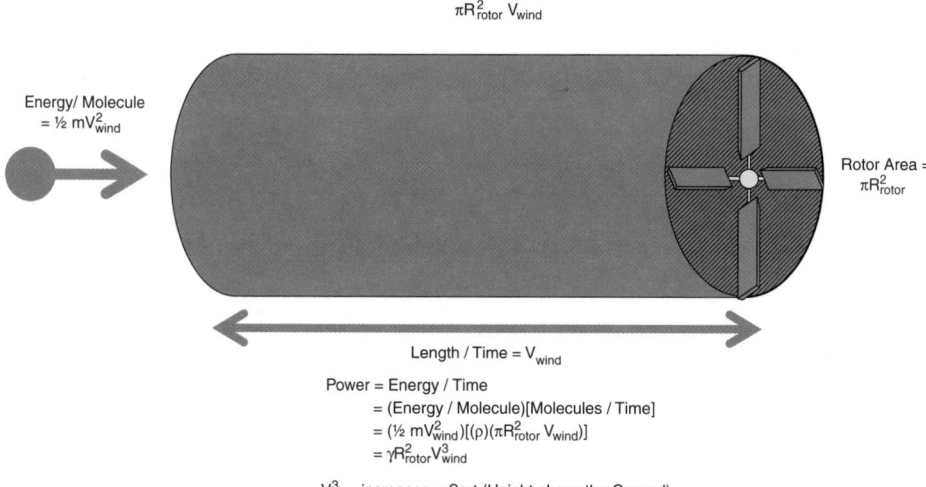

(Volume of Air Passing the Rotor Area) / Time =
$$\pi R_{rotor}^2 \, V_{wind}$$

Energy/ Molecule
= ½ mV_{wind}^2

Rotor Area =
πR_{rotor}^2

Length / Time = V_{wind}

Power = Energy / Time
= (Energy / Molecule)[Molecules / Time]
= (½ mV_{wind}^2)[(ρ)($\pi R_{rotor}^2 \, V_{wind}$)]
= $\gamma R_{rotor}^2 V_{wind}^3$

V_{wind}^3 increases ~ Sqrt (Height above the Ground)

Figure 4.11 Summary of the calculation of the power = energy/time delivered to a wind turbine indicating the scaling as the cube of the wind speed and the square of the rotor radius. In addition, at a given position on earth, the wind speed increases slowly with increases in height. A typical increase is (height/reference height)$^{1/7}$, but the actual increase in particular locations depends strongly on the topography of the location. For the case where the increase goes as the 1/7 power of the ratio, the cube of the wind speed increases approximately as (height/reference height)$^{3/7}$, which is approximately (height/reference height)$^{1/2}$ or the square root (sqrt) of the (height/reference height) above the ground.

Thus, the power that can be supplied by the wind is proportional to the product of the cube of the wind speed and the square of the rotor radius. The maximum power that actually can be harvested is approximately 2/3 of the maximum possible power calculated using this simple approach;[29] however, the proportionality relationship between the rotor radius and the wind speed is correct.

In the simplified calculation above, each air molecule is treated as a separate object whose motion is independent of the motion of any other

29. The actual energy contributed is slightly less, as discussed in online Appendix 6; however, the contribution still scales as the square of the wind speed, so the scaling law derived here is correct.

air molecule. Thus, it is assumed that the molecules behave like billiard balls that pass right by each other without any change in direction unless the balls actually touch. In reality, air molecules are not quite the individual billiard balls discussed above. Air molecules actually interact slightly with each other, which is another example of the action at a distance discussed in Chapter 2. In this case, the action at a distance is slightly attractive. The effect of the action at a distance that applies to air molecules can be felt by orienting the palm your left hand parallel to the floor face up. If the palm of your right hand is perpendicular to the floor and about 1 centimeter above the palm of the left hand, then moving the right hand back and forth can be felt as a breeze on the left palm. The feeling is stronger if the left palm is slightly damp. If air molecules behaved like billiard balls, one would not feel a breeze unless the air was directed toward the left palm. This interaction between air molecules has many important implications. For the case of wind power, it affects the ideal blade shape, as discussed in online Appendix 6.

The connection between air molecules also affects the scaling law governing the relationship between wind speed and height. At the surface of the earth the wind speed is zero as a result of the interaction between the air molecules and the stationary earth, which requires that the air molecules just above the earth be stationary; however, the wind speed increases rapidly with height, as anyone who has lain on a windy beach knows: one feels much less wind when stretched out luxuriously on a beach blanket than when standing up to walk to the snack bar. For wind power applications, the wind speed as a function of height can be expressed in terms of the wind speed at a given height. For example, if one chooses the standard height as 10 meters, at which the wind is $speed_{10}$, then the speed as a function of height is given by

$$speed(height) = speed_{10} \times (h/10)^{\alpha},$$

where h is the height in meters and α is called the Hellman coefficient. The values of the Helman coefficient vary with terrain. Typical Helman coefficients are 1/10 to 1/4, though significantly smaller and larger values are sometimes reported.[30] The typical Helman coefficients represented are a fairly weak dependence of wind speed on height, but the electrical power depends on the cube of the wind speed. Thus, even a weak dependence of wind speed on height can produce a significant increase in wind power with height, as illustrated in Figure 4.12, which corresponds to a Hellman coefficient of 1/7. The x-axis of the graph is relative height. If the relative height was chosen as 5 meters, then a relative height of 20 would correspond to $5 \times 20 = 100$ meters. Thus, increasing the rotor height from 5 meters to 100 meters would increase the generated wind power by a factor of three. This is why it is highly advantageous to position wind turbines on tall towers. This is also one of the reasons that it is extremely difficult to specify wind power density in general. When the NREL produces wind power maps, it always specifies the height at which the map applies. At the very same position on earth, the power density for a 100-meter tower is 3.5 times the density for a 5-meter tower. The trend toward taller towers is increasing the wind power density significantly.

Finally, there is the issue of how fast the rotor should rotate. Many people are puzzled that large wind turbines appear to turn very slowly, often less than one turn per second. An audience clapping every second would represent protest rather than adulation; however, for a large wind turbine even such a slow rate can represent quite a high speed at the tip of the rotor since the tip velocity is the product of the rotation rate times

30. See www.wind-power-program.com/UK_wind_speed_database; http://cdn.intech open.com/pdfs/17121/InTech-Methodologies_used_in_the_extrapolation_of_wind _speed_data_at_different_heights_and_its_impact_in_the_wind_energy_resource _assessment_in_a_region.pdf; www.byui.edu/Documents/physics/Theses/Bunker _Carolyn2010.pdf; http://ntrs.nasa.gov/archive/nasa/casi.ntrs.nasa.gov/19800005367 .pdf. "Modified Power Law Equations for Vertical Windspeed profiles D. A. Spera and T. R. Richard 1979 DOE/NASA/1059-79/4 NASA TM-79275; Raşit Ata and Numan S. Çetin. 2011. Analysis of height affect on average wind speed by ANN, *Mathematical and Computational Applications*, vol. 16, no. 3: 556–564, http://mcajournal.cbu.edu.tr/volume 16/vol16no3/v16no3p556.pdf.

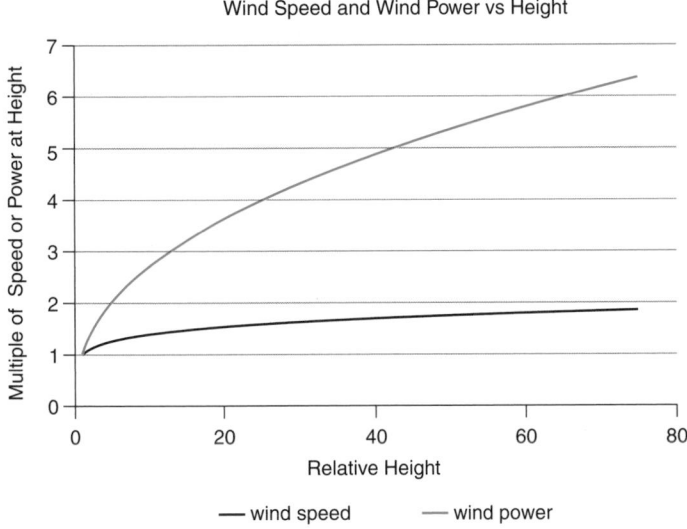

Figure 4.12 Relative wind speed and wind power as a function of relative height.

the radius. For a 5 megawatt wind turbine with around a 60-meter radius rotor, rotating once a second would result in a tip velocity of greater than 370 meters per second, which exceeds the speed of sound of about 340 meters per second. If the tip rotates at the speed of only 245 meters per second—the speed at which jet planes fly—then the turbine will complete a turn in 1.5 seconds. Thus, for large wind turbines, even slow rotation speeds correspond to extremely large tip velocities. Large turbines have specified tip speeds of 80 meters per second, which for a 5 megawatt turbine corresponds to around 1 rotation every five seconds.

The maximum possible energy that can be harvested from the wind is known as the Betz limit, which is derived in online Appendix 6. It states that for a given wind speed v, the maximum useable power is about 0.6 times the total power contained in the wind. Like combustion engines operating near the Carnot limit, present commercial wind turbines have efficiencies approaching the Betz limit, so significant efficiency increases are not possible; however, when fuel is burned to run a heat engine, the wasted energy must be extracted and paid for. In addition, the wasted burned fuel still produces emissions. As a result, for fossil fuel–burning systems, the lower the energy efficiency, the larger the negative environmental impact for a given amount of useful energy. In the case of wind turbine systems, the "wasted" wind simply blows past the turbine, which means that less efficient wind harvesting systems have lower environmental impacts.

I began this discussion of the basic science underlying wind power with the scaling laws that wind power is proportional to the square or the rotor length and the cube of the wind speed, where the wind speed increases with height. Thus, as a consequence of the basic science of wind power, large commercial wind farms located in windy locations that use large radius wind turbines positioned high above the surface of the earth will generate the largest possible power per unit of earth surface area.

Electrical Power Generation

Though wind power has vast potential, the power generated at any given time varies enormously because it scales as the cube of the wind speed. This enormously strong dependence of electrical power generation on wind speed is probably the dominant feature governing the feasibility of adopting wind power. This strong dependence of power on wind speed also affects where farms can be sited and creates new requirements for long-distance electricity transmission. In addition, it governs how much energy storage would be required to provide a constant reliable source of electricity from a wind power system.

A probability distribution for wind speeds is shown in Figure 4.13. For this particular distribution, the most probable wind speed is around 6 meters a second, but the maximum total power was generated during the much rarer times when the wind speed is about 12 meters a second. At that speed, the average power generated was a factor of three higher than the average power at the most probable wind speed even though the higher speed occurred only approximately one-third as often as the most probable

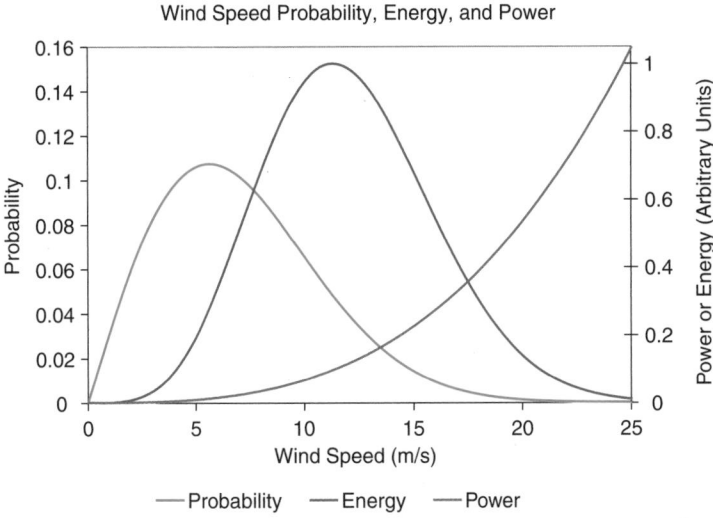

Figure 4.13 The probability distribution of wind speeds is shown in red. The corresponding electric power generation distribution is shown in purple, and the average energy demonstrated at each wind speed is shown in blue. Though the wind speed probability peaks at around 5 meters a second, the energy generated peaks when the wind speed is around 12 meters a second. This occurs because the energy scales as the cube of the wind speed, which allows higher wind velocities to contribute more electrical power than lower velocities, which have higher probabilities of occurring.

speed. This odd result is a consequence of the fact that the power generated is proportional to the cube of the wind. Thus, when the wind is blowing at 12 meters per second, it generates eight times as much power as it does when the wind speed is 6 meters per second. It generates eight times the power only one-third of the time that the wind speed of 6 meters per second generates power, so the ratio of the average power generated is ~8/3 (approximate), which is consistent with the graph.

Although average power is a useful measure if grid scale energy storage is available, in the absence of such storage one must consider how the availability of wind power changes with time. In order to provide enough power during lower-wind periods, the average power generated by windmills must be three to five times the average required power; however, when the wind speed is zero, no power is available, so even overbuilding capacity by a factor of one hundred cannot provide continuous power. At peak periods, the power generated can be more than thirty times the required power. Such huge fluctuations in power are hard to manage; however, commercial wind turbines are now designed to reduce the electric power fluctuations caused by changes in wind speed. They also include features to reduce the wear and tear on turbines by stopping power generation when the wind speed is below some chosen speed during times when the wind speed is so slow that very little electric power would be generated.

In order to discuss the performance of particular individual wind turbines, it is useful to understand the terms in Table 9 that characterize how the turbines are designed to respond to different wind speeds.

Thus, though mathematically the power generated increases with the cube of the wind speed, individual wind turbines are engineered so that the variation in wind power with the cube of the wind speed applies only between the cut-in wind speed and the cutout wind speed. For a given turbine, those speeds are chosen to minimize turbine wear with a minimum sacrifice in power production. Below the cut-in wind speed or above

Table 9.   Terms used to characterize the performance of individual wind turbines.

Wind turbine term	Meaning
Cut-in wind speed	Speed at which turbines start producing power, which may be significantly higher than the speed at which the turbine starts spinning
Rated wind speed	Speed at which the turbine produces the rated/nameplate power
Cutout wind speed	Speed at which the turbine stops producing power, though the turbine may still spin
Shutdown wind speed	Speed at which the turbine stops spinning to prevent damage
Survival wind speed	Speed the turbine is designed to withstand without falling over

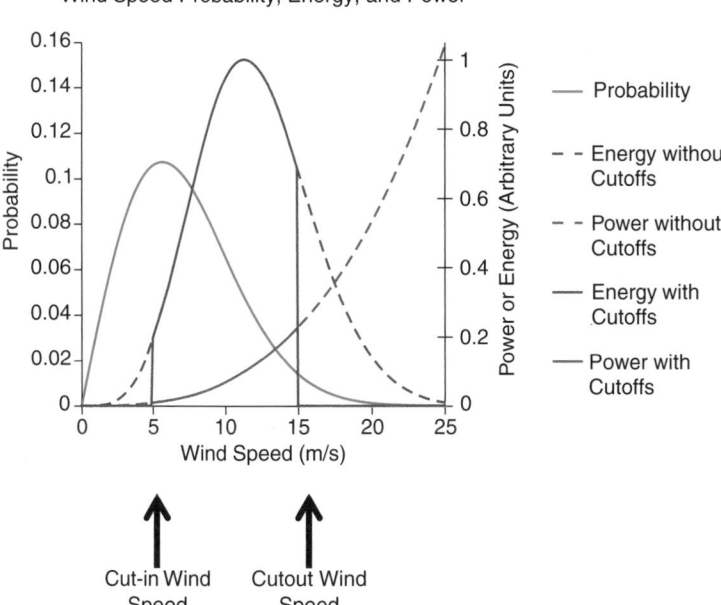

Figure 4.14 The effect of the cut-in speed and the cutoff speed on the power and energy generated as a function of wind speed. Below the cut-in speed and below the cutoff speed, the turbine generates no energy or power. For this example, the cut-in speed is shown as 5 meters per second. Typically it is lower, so it has a negligible effect on the energy generated. Similarly, the probability of high speeds is so low that despite the high wind power, the energy generated is very small. Thus, the cutoff may not decrease the total power generated significantly. The effect of the rated wind speed is not included.

the cutout speed, the turbine generates no power at all. The cut-in speed reduces wear during periods where very little energy would be supplied. Similarly, the cutout speed protects the generator, while the shutdown speed protects the rotor mechanism. Figure 4.14 is a graph of the effect of the cutout and cut-in powers on the energy and power generation.

As discussed above, many wind turbines not only have cutout speed, but they also manage power generation so that the power produced is limited to the power provided at some chosen wind speed. Above that

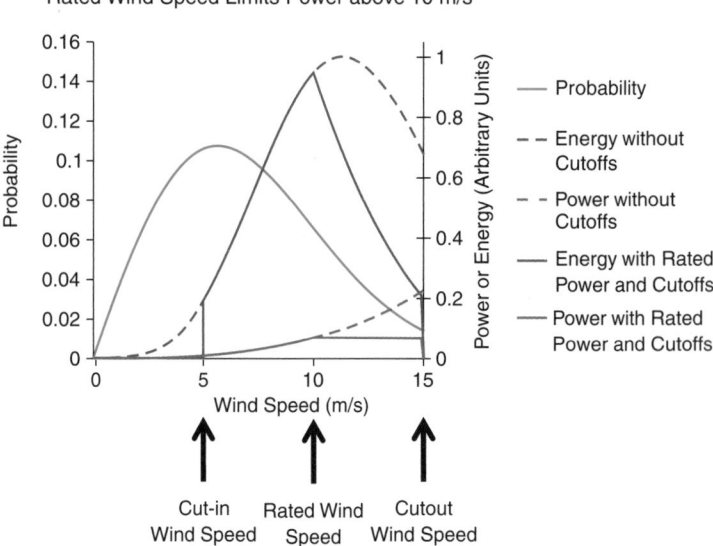

Figure 4.15 Illustration of the energy and power from a wind turbine that limits the maximum output power to the power achieved when the wind speed is 10 meters per second.

wind speed, the turbine does not generate more power, even though the wind could provide more power. This evens out the power generation but decreases the total amount of generated power.

The values of the cut in speed and the rated speed that are shown in Figure 4.15 were chosen in order to dramatize the effect of those values on the power generated. Thus, the chosen cut in speed is higher than typical and the chosen rated speed is lower than typical. Had typical values been shown, the effect of the cut in and rated speeds would be difficult to see, precisely because the cut in and rated speeds for operational wind turbines are selected so that they do not significantly reduce the generated wind power.

The fickle nature of the wind and the strong dependence of electrical power generation on the wind speed also mean that wind power capac-

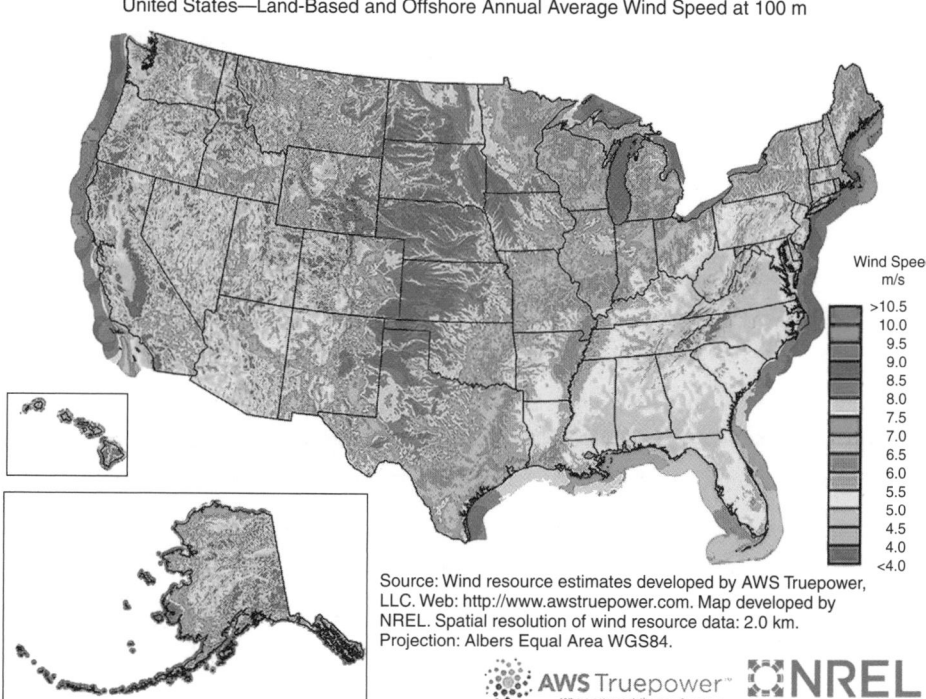

United States—Land-Based and Offshore Annual Average Wind Speed at 100 m

Source: Wind resource estimates developed by AWS Truepower, LLC. Web: http://www.awstruepower.com. Map developed by NREL. Spatial resolution of wind resource data: 2.0 km. Projection: Albers Equal Area WGS84.

Figure 4.16 National Renewable Energy Laboratory map of the United States showing wind speed at 100 meters. Blue corresponds to wind speeds greater than 10.5 meters per second, and wind speeds decrease in intervals of 0.5 meters per second until the lowest value of 4.0 meters per second, shown in green. As a consequence of the scaling of power with the cube of the wind speed, a place with a 10 meters per second average wind speed has two times the energy of a place with an average wind speed of 8 meters per second. Similarly, the wind power in the Southeast has only one-fourth of the wind power available in the Great Plains, which is still only one-half the wind power available off the coasts.

ity varies strongly from place to place, as shown in Figure 4.16. Thus, the siting of a wind power facility is more important than the siting of a fossil fuel–burning plant or a solar plant; however, a similar problem is encountered in the siting of hydroelectric dams since they require large water flows and geology suitable for damming. Figure 4.16 shows the average wind speed at various locations in the United States measured at

100 meters in height. Because of the scaling with height, a 10-meter tower would have about one-third the power density.

The highest average wind speeds are located well off the Northeast coast and fairly near the coast of northern California and southern Oregon. The next highest category is characteristic of the Northeast and Northwest coasts, as well as some regions in the northern Great Lakes. The Great Plains have average wind speeds similar to those found in many coastal regions.

The inland locations are often distant from the population centers where the majority of the U.S. electrical power is consumed. This separation between good wind power generation locations and the cities means that the generated power must be transported long distances. In contrast, for the last one hundred years, most electrical power consumed in the United States has been generated by burning hydrocarbons in plants that are very near to the cities that consume the power. Thus, long-distance transport of electrical power was not a significant issue previously but is becoming a very important issue as wind power becomes more important.

It is important to note that the average annual wind power shown in Figure 4.16 masks very substantial seasonal variations as indicated in Figure 4.17. In the absence of enormous energy storage, the large seasonal variations in wind power that result from the scaling of wind power with the cube of the wind speed will pose enormous challenges that might be missed if one assumes that the average values shown in Figure 4.16 are available year round.

As discussed above, Figure 4.16 shows that the largest wind speeds are located on the oceans, which is what motivates the creation of offshore wind power sites. Except for very mountainous regions, wind on land depends almost entirely on the weather. In contrast, oceans have a daily cycle of land and sea breezes that depend simply on the thermal storage capacity of the oceans. During the day, the oceans are cooler than the land, and during the night they are warmer. The difference in temperature results in a difference in pressure since lower temperatures are associated with lower pressures. This causes a pressure difference between the ocean and the land that is about 2 millibar, which results in a breeze blowing from the sea to the land during the day and from the land to the sea during the evening. The resulting wind speeds are approximately 8 meters per second at a height of 50 meters. Furthermore, the prevailing winds lie along a single direction, so using horizontal windmills with a fixed axis direction does not significantly reduce efficiency. The advantages of offshore locations are driving wind power offshore; however, at present, the levelized cost of offshore wind (22.1 cents per kilowatt-hour) still exceeds the cost of onshore wind by (8.66 cents per kilowatt-hour) by a factor of approximately two according to the levelized cost estimates from the EIA.

Figure 4.18 shows the nameplate wind power capacity installed in different states. If, consistent with EIA estimates, one assumes that the average power is around one-third of the rated power, then one can convert the nameplate capacity shown in Figure 4.18 into average available wind power by multiplying by 0.3. Broadly, the installed wind power shown in Figure 4.18 illustrates the seasonally averaged available wind power, indicating that people prefer to locate wind farms in windy places; however, consumer density and politics also play a role, as indicated by New York, which has more installed wind energy that the wind speed map would suggest.

In general, the map indicates that wind power is often located far from population centers, with very little power produced along the very high population density "Acela corridor" that runs from Boston to Washington. Transporting power long distances results in energy losses due to the electricity traveling along the wires. These losses have various sources, where different loss mechanisms require different mitigation strategies. In general, losses are reduced by increasing the voltage at which the power is delivered, as discussed in Chapter 7, which also considers the issue of whether the power should be transported as AC power or DC power.

Given the challenges in transporting and converting power, one might imagine that local small-scale windmills would be more cost effective;

SPRING

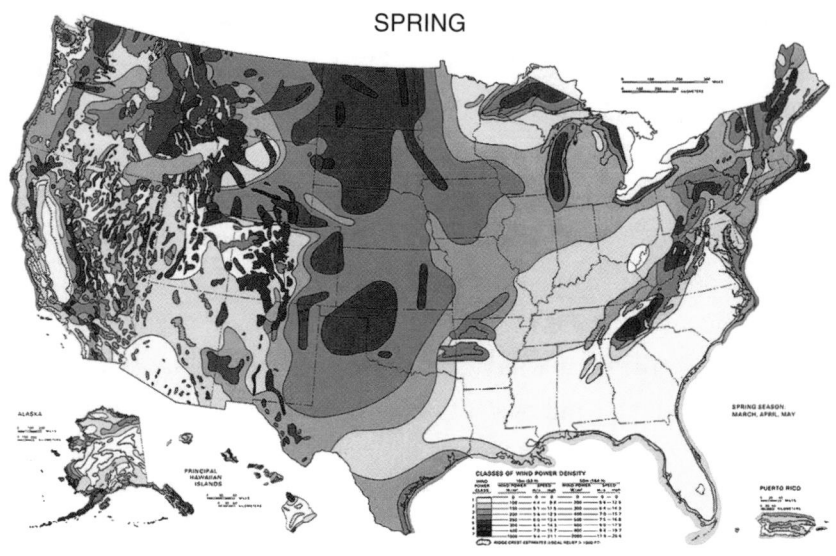

AUTUMN

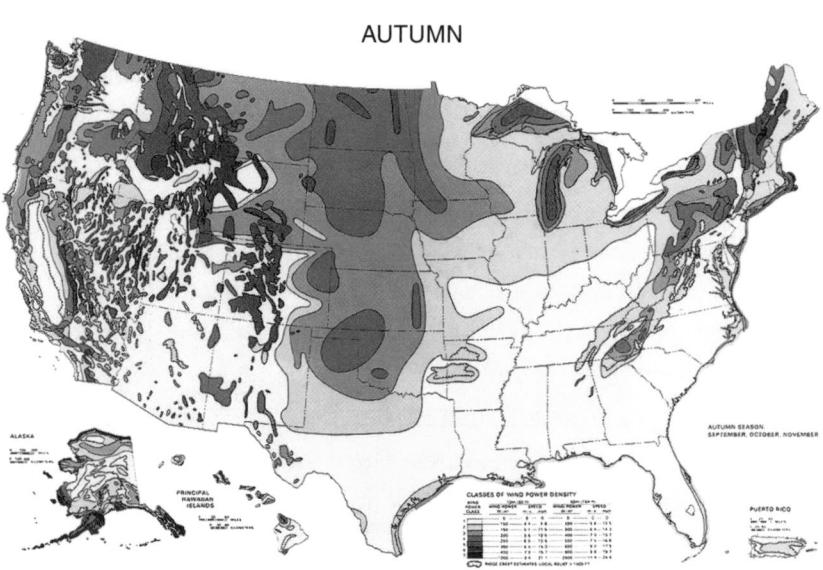

Figure 4.17 National Renewable Energy Laboratory map of the seasonal average wind speed in the United States, where darker blue indicates higher wind speed. Clockwise from the upper left the seasons are spring, summer, winter, and fall. The maps show that in most of the United States, wind speeds

SUMMER

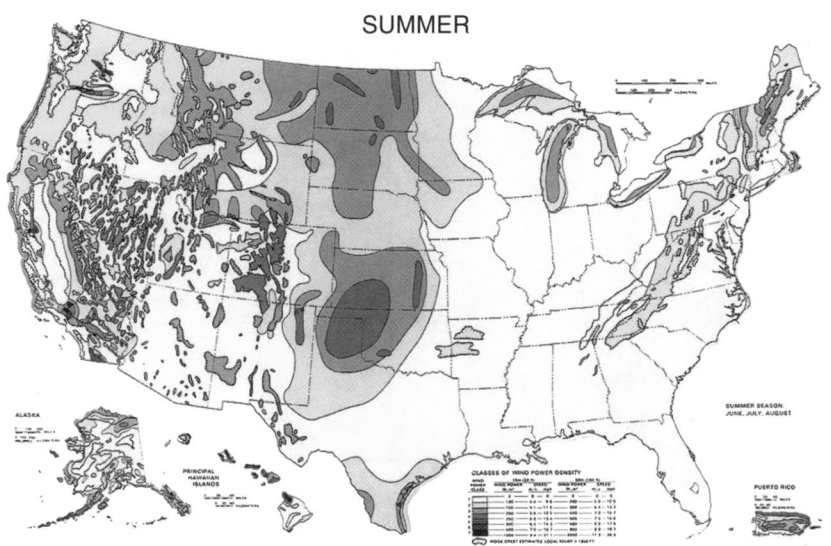

WINTER

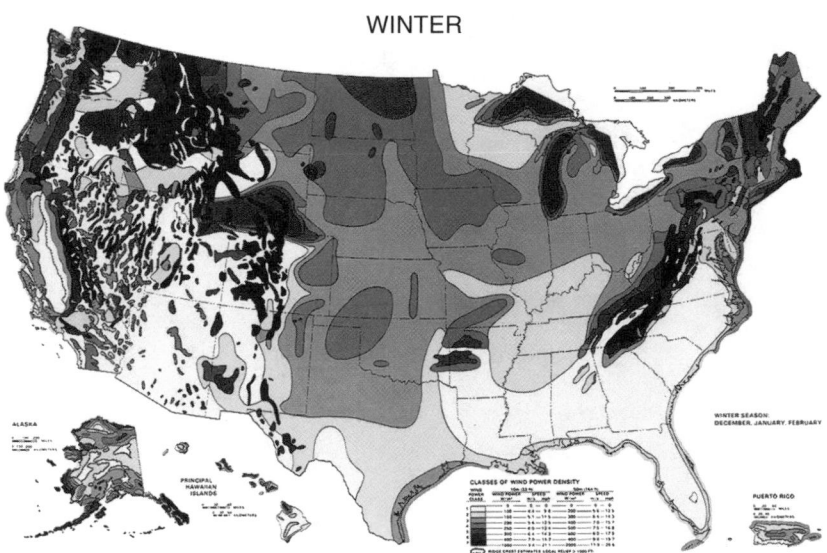

in the winter and spring are substantially larger than the wind speeds during the fall and summer; however, areas of California are an important exception as some areas, such as San Francisco, have wind speeds that peak in May and June, though the wind in Los Angeles peaks in April.

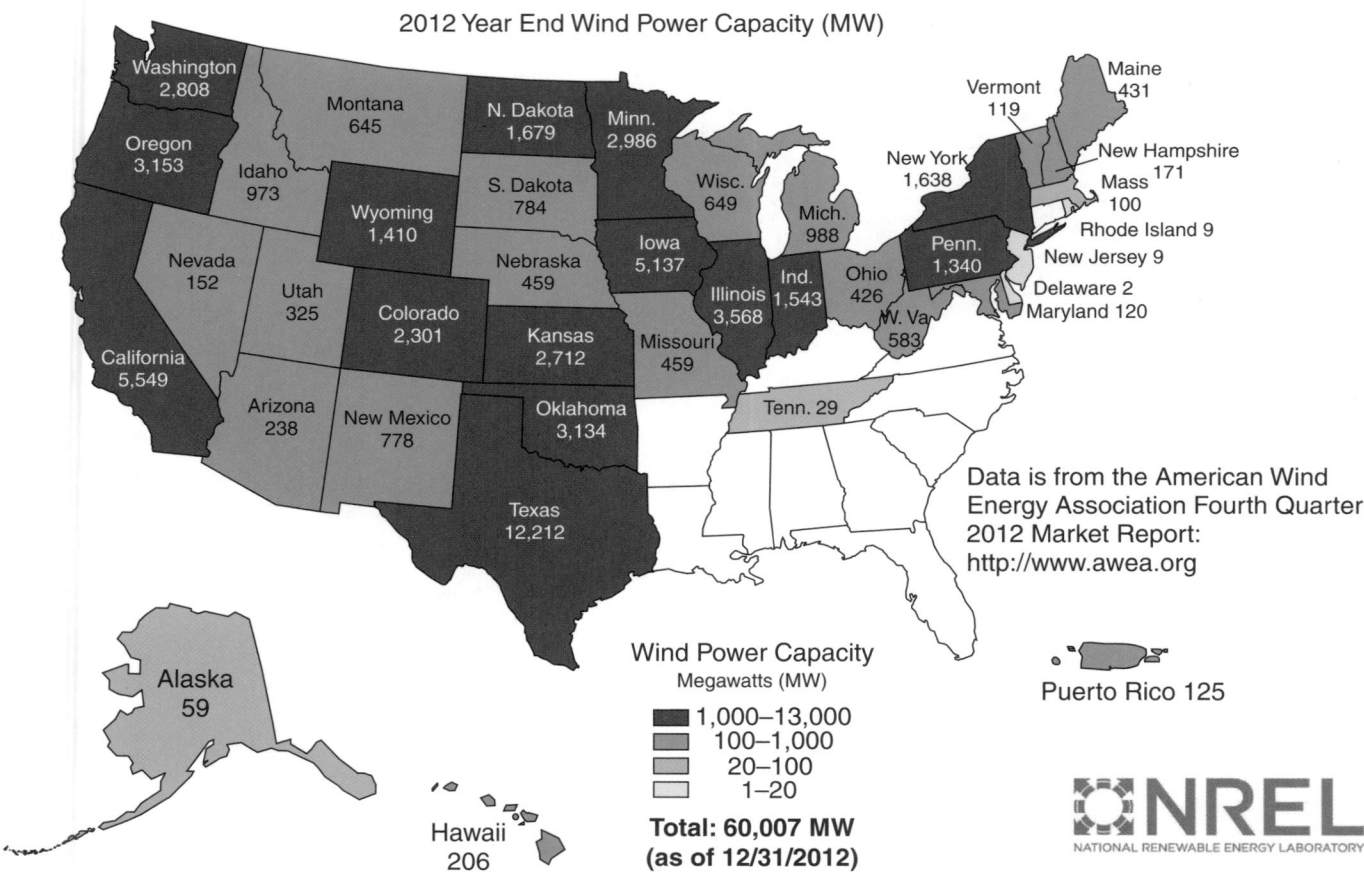

2012 Year End Wind Power Capacity (MW)

Washington 2,808
Montana 645
N. Dakota 1,679
Minn. 2,986
Vermont 119
Maine 431

Oregon 3,153
Idaho 973
S. Dakota 784
Wisc. 649
New York 1,638
New Hampshire 171

Wyoming 1,410
Mich. 988
Penn. 1,340
Mass 100

Nevada 152
Utah 325
Nebraska 459
Iowa 5,137
Illinois 3,568
Ind. 1,543
Ohio 426
Rhode Island 9
New Jersey 9

California 5,549
Colorado 2,301
Kansas 2,712
Missouri 459
W. Va 583
Delaware 2
Maryland 120

Arizona 238
New Mexico 778
Oklahoma 3,134
Tenn. 29

Texas 12,212

Data is from the American Wind Energy Association Fourth Quarter 2012 Market Report: http://www.awea.org

Wind Power Capacity
Megawatts (MW)
1,000–13,000
100–1,000
20–100
1–20

Alaska 59

Hawaii 206

Puerto Rico 125

Total: 60,007 MW (as of 12/31/2012)

NREL
NATIONAL RENEWABLE ENERGY LABORATORY

Figure 4.18 State-by-state map of the installed wind power in the United States. Source: National Renewable Energy Laboratory.

however, in the United States, transmission losses are only about 6 percent. In contrast, the wind power generated is proportional to the square of the length of the rotor and the cube of the wind speed. In addition, the wind speed increases with height. As a result, the loss of efficiency due to scaling down the windmill size may overwhelm the gain associated with shorter transmission distances. Thus, it is beneficial to construct windmills with large rotors on the top of enormous towers in very windy locations. Many people find such large towers aesthetically displeasing, so the siting of wind power includes an additional non-engineering-based factor: the local opposition to installing windmills that "deface" the landscape. This has been a contentious issue in Massachusetts, where the Cape Wind project, which plans to locate a large wind farm just off the coast of Cape Cod, has generated enormous opposition from local homeowners with views of the Nantucket Sound. The political tension in this case has also been increased because many of the home owners are politically influential. Ted Kennedy was one of the most liberal American politicians in the twentieth century, but he opposed the Cape Wind project because the windmills would mar the view from his family compound in Hyannis. Despite politics, science favors siting wind power on Cape Cod because the average wind speeds are so large, as shown in Figure 4.16. Even in the summer, the winds blowing on the Cape are significant, whereas most of the rest of the Northeast is nearly becalmed, as shown in Figure 4.17.

Despite the many challenges associated with wind power, the rate at which wind power capacity has been increasing is enormous, as we saw in Figure 2.7. This increase is occurring in both the developing and the developed world. In 2011, the wind power–generating capacity in China exceeded the wind power–generating capacity in the United States for the first time. Practical large-scale energy storage would greatly improve the economics and practicality of wind-based power.

WIND POWER ON AVERAGE CAN PROVIDE
100 PERCENT OF U.S. ENERGY

A summary of the average power that wind could provide is given in Table 10. Similar information for hydroelectric and photovoltaic solar power is shown in Tables 8 and 11, respectively; however, Table 8 lacks a third column because more than 100 percent of U.S. land area would be required in order for hydroelectric power to meet even the current electrical power demand for the United States.

Table 10 shows that wind power can easily meet the average energy needs for the United States; however, due to the fluctuating nature of wind power, it would be extremely difficult and expensive to meet the instantaneous power demands of the United States even though periods of zero power generation are infrequent. Current wind turbines already include features to reduce the variability of power generation despite the fundamental variation in wind speed. Using a turbine that levels power at wind speeds above the rated speed can greatly reduce power fluctuations, but that reduction in fluctuation is accompanied by a reduction in the total power generated. This is an efficiency issue. It means that when the wind speed exceeds the rated speed, no more electrical power is generated even though the wind offers substantially more power.

Demand can be met more constantly if the maximum wind turbine generation is much larger than the average demand, but that means that the wind turbines are often generating too much power. At higher wind speed, some turbines can simply be turned off, which allows the remaining turbines to meet the demand; however, turning off most of the turbines is an economically wasteful option. Having sufficient on-site energy storage to contain the extra generated power is challenging and expensive. An alternative is to use the excess power for a purpose like desalination or charging of electric car batteries. Such applications do not require continuous power, so they allow all turbines to continue to exploit all available power at high wind during times when the resulting electricity supply

Table 10. Feasibility of wind power for providing all U.S. energy needs or just the country's electrical energy needs.

Quantity characteristic wind power in the United States	Value	Percent of U.S. land area for 100% total power
5-MW generator has 130-m diameter	n.a.	n.a.
Cost/peak watt	$1	n.a.
Cost/average watt at 30% capacity factor (DOE estimates $2.40)	$3	n.a.
Assuming 15 times diameter spacing gives $\pi[(D*15)/2]^2$ area	$2.99 \times 10^{6\,06}$	n.a.
Total U.S. power/area = 5 MW/area 30% capacity factor	0.50	64.9%
Assuming 10 times diameter spacing gives $\pi[(D*10)/2]^2$ area	$1.33 \times 10^6 E$	n.a.
Total U.S. power/area = 5 MW/area 30% capacity factor	1.13	28.8%
Assuming 7 times diameter spacing gives $\pi[D*7)/2]^2$ area	7.33×10^5	n.a.
Total U.S. power/area = 5 MW/area 30% capacity factor	4.09	8.0%
Assuming 3 times diameter spacing gives $\pi[D*3)/2]^2$ area	1.19×10^5	n.a.
Total U.S. power/area = 5 MW/area 30% capacity factor	12.56	2.6%
Number of wind turbines required for total energy 30% capacity factor	2.1×10^6	n.a.
Number of wind turbines required for electric energy 30% capacity factor	2.9×10^5	n.a.
Cost of wind turbines required for total energy 30% capacity factor	9.6×10^{12}	n.a.
Cost of wind turbines required for total energy excluding thermal losses at 30% capacity factor	$5. \times 10^{12}$	n.a.
Cost of wind turbines required for electric energy 30% capacity factor	1.3×10^{12}	n.a.
Cost of wind turbines required for total energy 30% capacity factor/GDP	61%	n.a.
Cost of wind turbines required for total energy after losses 30% capacity factor/GDP	31%	n.a.
Cost of wind turbines required for electric energy 30% capacity factor	8%	n.a.

Note: Various power/area scenarios are considered, where each has a different characteristic spacing between turbines, using 5MW commercial turbines to estimate relate nameplate power to area. The percentage of U.S. land area is calculated only for 100 percent total energy. To get energy after losses, divide that number by 2. To get percent required for electricity only, multiply by 0.13.

exceeds the immediate demand. In the desalination application, the expense of the water must include the time that the physical plant is idle because the wind speed is too low. In contrast, in the battery charging scheme, there is no significant physical plant standing idle when the wind power is low. Such a scheme could allow battery-powered cars to change battery packs at service stations, a process that might be much faster than charging the batteries in the car.

5

ELECTRICITY FROM THE SUN

In contrast to systems that use a rotating shaft to generate electricity, photovoltaic solar power does not require any moving parts at all. In such systems, sunlight is transformed directly to electrical power. The absence of moving parts greatly reduces system complexity as well as maintenance requirements. If the solar cell materials did not degrade with time, a solar cell would produce electricity forever. Existing commercial solar cells can produce electricity for twenty to thirty years before the materials degradation is significant. The longevity and simplicity of such systems is enormously attractive, but unlike hydroelectric power and windpower, which indirectly harvest sunlight, solar PV systems directly harvest sunlight. Thus, though animals can graze and crops can be grown on land "used" for hydroelectric power or wind power, the area used for solar cells is otherwise completely dark, limiting crop growth to fungi. As a result, photovoltaic solar (PV) systems are most desirable in locations where the sunlight serves no immediate human purpose. Building roofs and deserts are two such locations. In the developing world, solar PV can make enormous contributions by providing lighting and electric power to regions of the developing world where grid-delivered electricity is unavailable, unreliable, or prohibitively expensive.

OVERVIEW OF PHOTOVOLTAIC SOLAR ELECTRICITY GENERATING SYSTEMS

In photovoltaic solar (PV) systems, sunlight is converted directly into electrical power, and it is theoretically possible for systems to approach 100 percent efficiency. Solar power numbers are very confusing. A quick Google search provides many sources stating that the sun delivers 1,365 watts per square meter to the earth, and the Solar Radiation and Climate Experiment (SORCE) experiments measure approximately 1,361 watts per m^2 at the top of the earth's atmosphere facing the sun.[1] Of course, no place on earth always faces the sun. Thus, because of the rotation of the earth, the number 1,365 must be divided by two to take into account the times during which any particular position on the earth is in darkness. Even at the equator, just as the sun is rising the sun's rays are almost parallel to the earth. Thus, a given square meter of sunlight is deposited over a large area, which is proportional to $1/\cos(\theta)$, where θ is the angle between the a line perpendicular to the earth's surface and the center of the sun, as illustrated in Figure 5.1. Finally, clouds and other material in the earth's atmosphere absorb some sunlight. As a result, the average power delivered to the earth by the sun is much smaller than the 1,365 watts per square meter delivered to a satellite high above the earth. Even in the absence of atmospheric absorption, the two effects associated with the earth not facing the sun will reduce the average power by a factor of one-half due to time facing away from the sun and $2/\pi$ due to the angular effect, reducing 1,365 to 430 watts per square meter, Solar panel inefficiencies make the delivered electrical power lower than the power delivered by the sun by another factor of approximately four.

The decrease in sunlight over area that arises at the equator each day as a result of the rotation of the earth is similar to the decrease in sun-

1. See http://lasp.colorado.edu/data/sorce/total_solar_irradiance_plots/images/tim_level3_tsi_24hour_640x480.png.

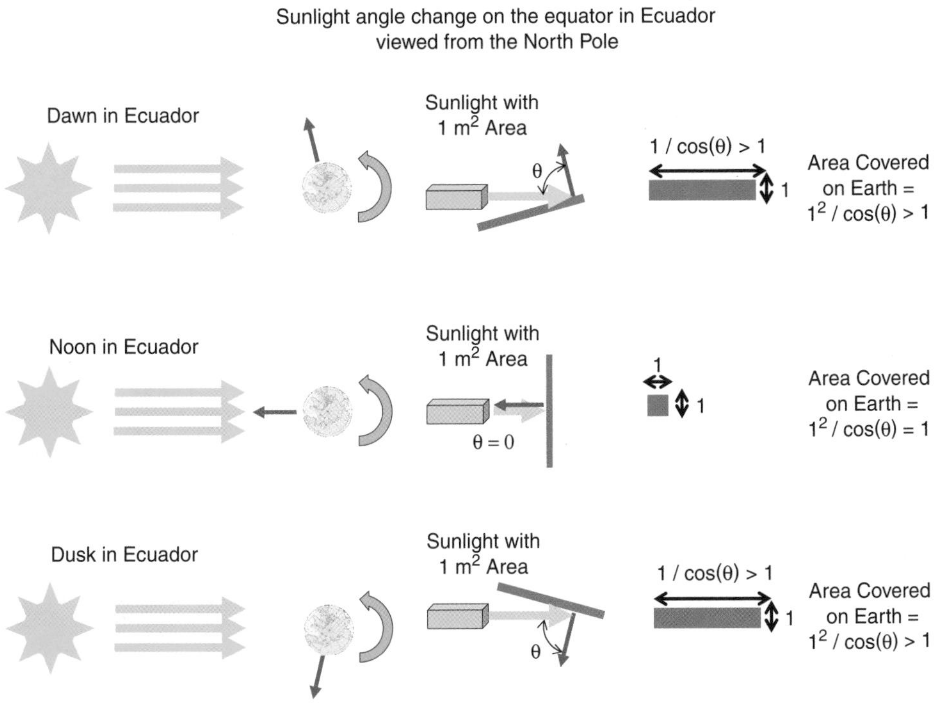

Figure 5.1 Illustration of why the sunlight over the area of earth's surface is largest at noon. The drawing shows that the sunlight that would cover 1 square meter at noon is spread over $1/\cos(\theta)$ square meter on the surface of the earth, where theta is the angle between the incoming sunlight and the arrow pointing in the direction perpendicular to the earth's surface. The green rectangles show the area on the earth's surface that is hit by the 1 square meter of sunlight. This is the same effect that causes shadows to be shortest at noon and very long near dawn or dusk. Image of the world from the North Pole provided by the CIA World Factbook.

light over area that arises at higher latitudes even at noon. At the equator, the sun light is almost perpendicular to the surface of the earth at noon; however, farther north, even at noon the rays of the sun are not perpendicular to the earth. As a consequence, the sunlight over area delivered to the surface of the earth is lower at higher latitudes, as shown in Figure 5.2. The two rectangular shapes contain an equal amount of energy delivered from the sun. At the equator, the sunlight is nearly perpendicular to the earth, so all of the sunlight falls onto a square with area d^2. In

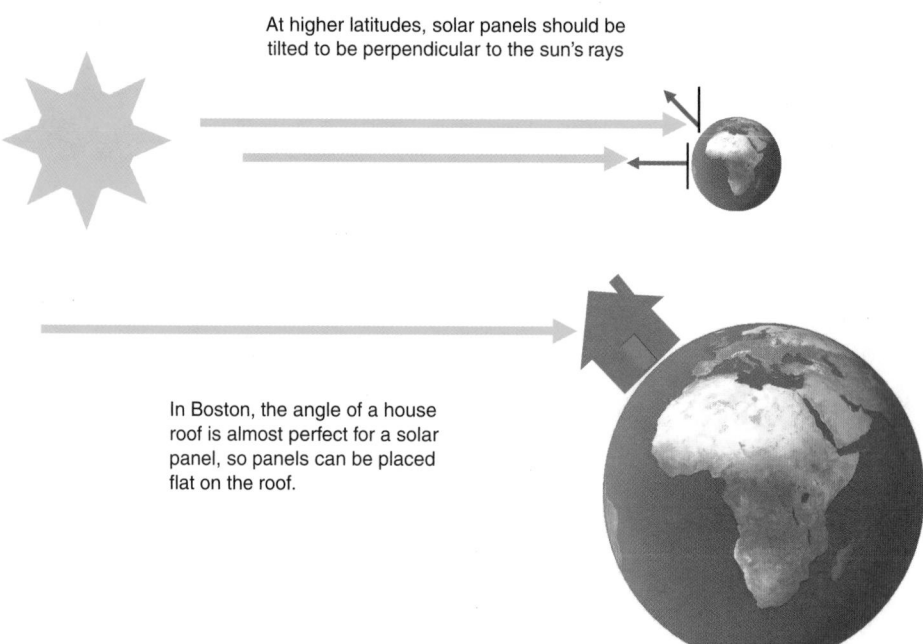

Figure 5.2 Illustration of the effect of latitude on the amount of sunlight hitting one square meter on the surface of the earth. At the equator at noon, the sun's rays are perpendicular to the earth; however, even at noon in Boston, the sun's rays are tilted by approximately 45 degrees. At higher latitudes, tilting south-facing solar panels so that they are perpendicular to the sun's rays at noon removes the angular loss factor resulting from latitude, which is why solar panels in the United States are usually tilted. Conveniently, for roofs with a central axis running east to west, the angle of the roof is approximately the ideal angle for solar panels, allowing them to be placed directly on the surface of the roof. Image of the earth from the National Aeronautics and Space Administration by Reto Stöckli and Robert Simmon.

contrast, away from the equator the sunlight hits the earth at an angle, so the area on the surface of the earth that intercepts the same amount of sunlight is larger by a factor of $1/\cos(\theta)$, where θ is the angle between the sun's rays and the perpendicular to the surface of the earth. Thus, less sunlight falls over unit area if the solar panel is parallel to the surface of the earth. Of course, one can compensate for this effect by placing the solar panels at angles, which fortunately are approximately equal to roof angles for people in Europe and the United States, with rooflines running from east to west.[2]

Figure 5.3 is a map from the NREL showing the average solar power available in the United States assuming that the solar panel is tilted at the appropriate angle for the latitude. Red regions have more available power. The maximum value of around 6.6 kilowatt-hours per square meter a day occurs in the southwestern deserts, with around 4 kilowatt-hours per square meter a day being typical in the Northeast. While the map uses kilowatt-hours per square meter a day, it is convenient to reexpress these values in terms of average watts per square meter, so 6.6 kilowatt-hours per square meter corresponds to around 277 watts per square meter of average power for the Southwest, which is only slightly smaller than the theoretical maximum of approximately 430 watts per square meter that I calculated above based on the 1,365 watts per square meter flux falling on a satellite in orbit without including any losses traversing earth's atmosphere. The solar situation in cloudy Boston is worse since the of average power for the Northeast is only around 170 watts per square meter. Rainy and cloudy Seattle has an even lower flux. Thus, in the United States, solar energy delivery varies by a factor of around two, just as wind speed does; however, since wind power scales as the cube of the wind speed, the factor of two variation in wind speed translates to approximately a factor of ten in electricity generated, whereas the factor of two variation in solar energy corresponds to only a factor of two variations in electricity generator. As a result, the siting of wind farms is much more important than the siting of solar panels.

In practice, photovoltaic solar (PV) systems convert about 10 to 25 percent of sunlight into electrical energy, though research systems have demonstrated much higher efficiencies. According to the EIA, in Arizona

---

2. Thus, solar panel placement of roofs is both convenient and effective. Similarly, one could reduce the effect due to the rotation of the earth by rotating the solar panels so that they face the sun; however, that requires energy and greatly complicates the solar power system, which is not worthwhile for the factor of about two that one would gain in energy.

United States Photovoltaic Resource: Flat Plate Tilted at Latitude

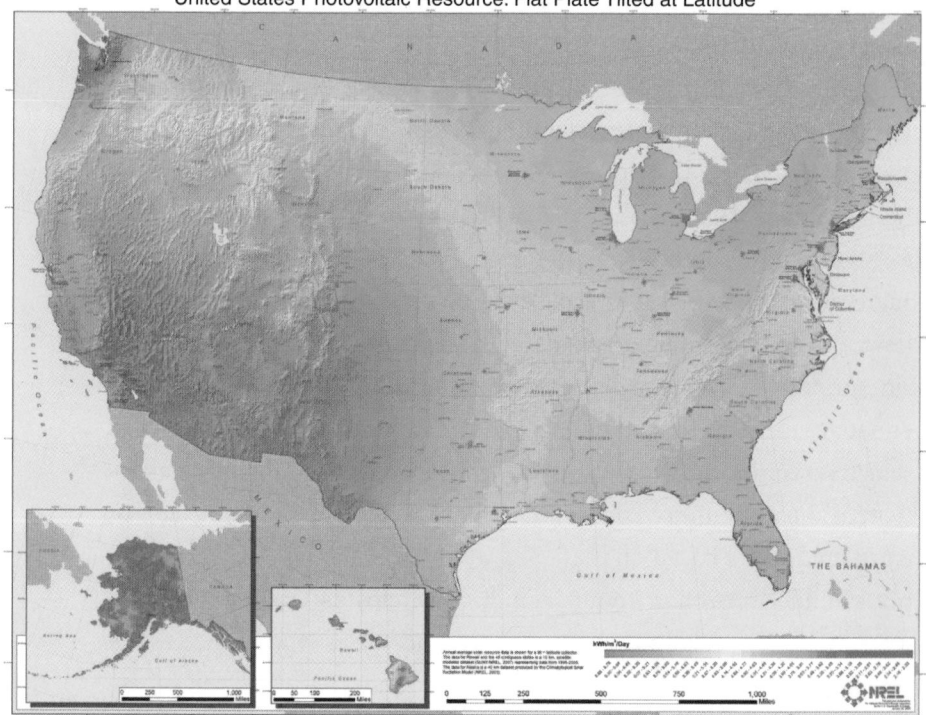

Figure 5.3 Insolation map from the National Renewable Energy Laboratory showing the average kilowatt-hours of sunlight per day available in different regions of the United States. In general, hotter, sunnier areas have greater average solar flux. Red corresponds to an average power of 277 watts per square meter, and green corresponds to an average power of around 166 watts per square meter. The difference between red and green regions is less than a factor of two. In contrast, for the corresponding wind power map, the difference was more than a factor of eight, highlighting why the siting of wind power plants is much more crucial than the siting of solar panels.

the electricity consumption per household is 1.3 kilowatts, and in Massachusetts it is about 880 watts. Thus, if the panels were 100 percent efficient in Arizona, 3 to 4 square meters of solar panels would provide the average power for a household, whereas in Massachusetts 5 to 6 square meters of solar panels would be required; however, solar panels are not 100 percent efficient. Given typical solar panel efficiencies of around

20 percent, the actual required areas are five times larger than the areas calculated for 100 percent efficiency. Even at 20 percent efficiency, solar panels on the roof of detached single-family homes should be able to generate the average power consumed by the household, as I will show next.

The average house area is around 2,500 square feet, so assuming that space is divided between two stories, the average surface area of the roof would be around 1,000 square feet, which is approximately 100 square meters, producing an average power in excess of 40 W/m$^2 \times$ 100 m$^2 =$ 4 kW. If each dwelling used one-half the roof area for solar panels, then on average each house generate two to three times as much electricity as it uses. In other words, each of the 70,000,000 single-family houses in the United States would generate 2 kilowatts of power, or around 1.4×10^{11} watts, which is approximately one-third of the total electrical power used in the United States.[3]

Confirmation that this idea is not fanciful comes from Germany, where on Saturday May 26, 2012, solar energy provided over half of the electricity consumption in Germany at midday and 20 percent for the twenty-four-hour day. It is important to note that May 26, 2012, was a sunny weekend day near the summer solstice and that Germany is located at a fairly high latitude, resulting in more than 12 hours of sunlight in midsummer. The fractions on dark winter weekdays are much smaller; however, on average solar power provides around 5 percent of Germany's electrical energy, which is comparable to the U.S. number for hydroelectric power.

Of course, technical feasibility does not mean that the switch to solar would be economically feasible. I will consider the simple case of putting a solar panel on every single-family detached house in the United States. In Massachusetts, people pay around $0.15 per kilowatt–hour, whereas

---

3. On average, half of the roofs have a north-south axis instead of an east-west axis, so the solar panels could not be placed directly on the roof but would have to be placed at an angle.

in Arizona they pay only around $0.11 per kilowatt-hour. Home Depot sells solar panel kits for $1–$2 per peak watt, where average watts are approximately one-quarter of peak watts since on average it is dark half of the time and even in the light the solar panels face the sun only at noon. Thus, 1 kilowatt of average consumption would require 4 kilowatts of peak generation, which with installation costs between $5,000 and $10,000 dollars; therefore, based on Massachusetts or Arizona electrical rates, the capital expense would be paid back in eight to ten years. Since solar panels are expected to last twenty to thirty years, the invested money would return a profit of around 200 percent over twenty years, which corresponds to between 3 and 5 percent interest. Thus, putting solar panels on every roof is economically reasonable but not a path to riches.

Putting a solar panel on the roof of every single-family home represents a significantly different paradigm of energy investment and consumption in which the energy is provided directly by the consumer to himself, rather than by a utility to which the consumer pays money. Of course, one need not confine such systems to single-family homes. Big box stores, shopping malls, schools, factories, and houses of worship also have unused roof space that might reasonably be outfitted with solar panels whose output would primarily be used by the owner of the panel. Ikea uses solar panels on the roofs of its stores and sells solar panels to its customers. Lacking effective local energy storage, consumers usually sell the excess electricity generated during the day to power companies, and then during the night consumers draw power back from utilities that generate electricity in large power plants. Each consumer then also acts as a local provider who adds energy to the grid. Given that air conditioning is the number one source of electricity use in the United States, the ability to generate peak solar power during peak air conditioning periods may substantially lower overall energy prices by reducing capital that must otherwise be tied up in rarely used physical plants. Oddly, the reduction in energy prices for everyone lowers the rate of return for those who invest in the solar panels. Such subtle factors are often neglected in energy cost analyses.

THE SCIENCE UNDERLYING PHOTOVOLTAIC SOLAR

In photovoltaic systems, energy from the sun generates electricity by interacting with a photovoltaic material, such as a silicon photodiode. The basic principle underlying a silicon photodiode is illustrated in Figure 5.4. The photodiode contains a region, known as the depletion zone, characterized by a very strong electrical potential hill. If there is a free charge in this hill region, the charge will gain energy by falling down the hill. In the absence of light, there are no free charges on the hill. When light hits the photodiode, the light provides energy that frees charges that were trapped in the hill region. Once the charges are freed in the hill regions, they gain energy by falling down the hill. Thus, the hill is always present, but free charges occur only when the hill of sunlight comes in and frees them. This system is illustrated in Figure 5.4 along with a gravitational analogy.

First, consider the gravitational system illustrated by a hypothetical hydroelectric system based on a desert mesa. The top of the mesa is flat, and the bottom of the mesa is flat, but there is a slope in between. At the surface of the earth, an amount of water with a mass m is pulled toward the center of the earth by a gravitational force given by mg. If that water falls down a mesa with a height h, the energy of the water at the bottom will be mgh, if no energy is lost to friction. Mesas tend to occur in the desert, so there is no available water on the surface; therefore, despite the strong gravitational hill, there is usually no water falling down the mesa. Thus, desert mesas are not good locations for hydroelectric power plants, but deep inside the mesa, there can be wells or holes in the earth that contain water. If a person were to haul buckets of water up from the well and dump the water onto the slope of the mesa, the gravitational force would then pull the water down the hill. Thus, the falling water could do mechanical work once it has been lifted out of the hole and released

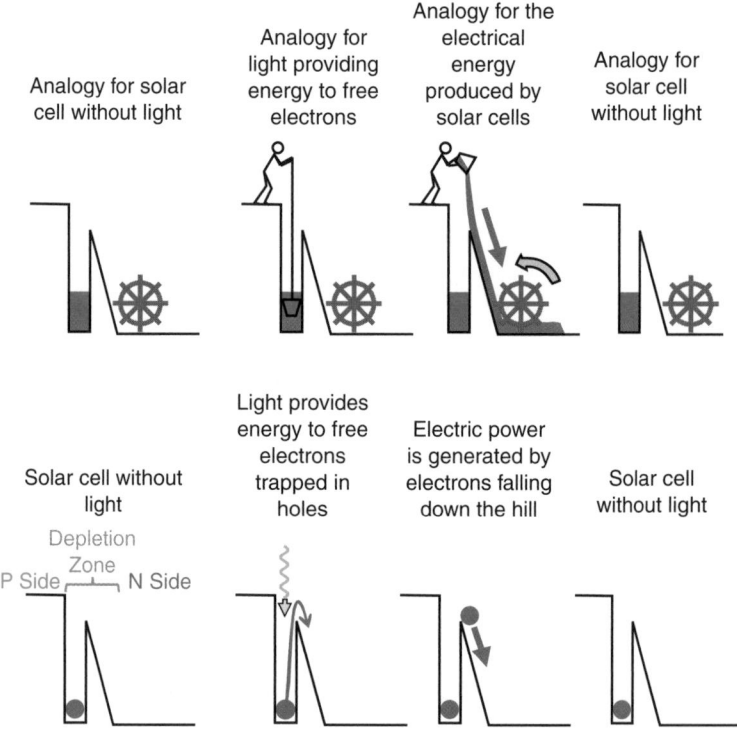

Figure 5.4 Illustration of a solar panel and its gravitational analogy. The upper illustrations show the analogy, while the lower illustrations show the solar cell. In both cases, the first panel indicates the case where there is no light. Even without light, there is a strong potential hill. In the gravitational analogy, there is no free water on the hill, only water trapped in wells. In the solar cell, there are no free electrons, only electrons trapped in holes. Thus, no water or charge flows. The second panel illustrates how light falling on a solar cell provides energy to create free charge by lifting electrons out of holes, which in the gravitational analogy is represented by the work done by a person lifting water out of a well. The third panel shows the freed water falling down the gravitational hill, just as the freed electrons fall down the electrical hill. In the analogy, the energy gained by the water can do work that makes the wheel turn, and in the solar cell, the energy gained by the electrons can be used to do work. The final panel is the same as the initial panel. It represents what happens if the light is turned off after it has been on for awhile. The supply of electrons in the holes is constantly replenished, so there are always trapped electrons available for the photons to free.

onto the side of the mesa. Water will continue to flow and be available to do work as long as a person continues to bring water up from the well and release the water down the side of the mountain. If the person stops, the water flow will stop. Thus, in the absence of a person who brings water to the surface, no gravitational energy can be obtained from the desert mesa even though the downward slope of the mesa is always present.

The mechanical example is rather silly, since the work done to raise the water from the well is the same as the work that the water can do when it falls back down; however, in solar cells this example does make sense because in the solar power system, the energy to lift an electron out of a hole comes from the sunlight hitting the photovoltaic cells.

The electrical system in a solar voltaic cell is completely analogous to the hypothetical mesa system. Inside a so-called PN or PIN junction solar cell, there is a built-in electrical potential mesa that is part of the design of the device and which remains in the device forever. The PN junction is located at the junction between the P-doped region and the N-doped region. At the junction between the P and N regions, there is a depletion zone. In the absence of light, the depletion zone contains no free charge. The free electrons that were in the depletion zone have all fallen down into holes in the depletion zone, where they have become trapped; however, if light is shining on the junction, the incoming light can provide enough energy to lift some of the electrons out of the holes. In drawings, photons (light particles) are usually represented by squiggly lines, as they are in Figure 5.4. Once the light lifts an electron out of a hole, the electron is free to roll down the potential well. The current resulting from those falling electrons can power electrical devices. The energetic depths of the holes is about the same as the height of the hill, so the conversion of light energy to electrical energy can be very efficient.

In hydroelectric systems or a wind power system, the kinetic energy of the flowing water or air has to be converted to kinetic energy in a rotating generator that in turn has to be converted into electrical current. In contrast, in a photovoltaic cell, the energy from the sun is converted directly into a moving charge. No moving parts are required. Sunlight simply shines on the appropriately constructed piece of silicon, and electrical power emerges. The power continues to emerge as long as the sun is shining. If the sun stops shining, then there will be no free charges to fall down the hill. Thus, no electrical power will be produced even though the electrical hill is always present in the silicon. The process that converts sunlight to electrical energy can be more than 90 percent energy efficient for photons of the right color that are not reflected from the silicon.

Though the theoretical efficiency of a solar cell is very high, the practical efficiency is not. One source of energy loss in a solar cell is reflection of the light from the silicon back toward the sun. Most solar panels are shiny. If you can see yourself, that means that light hits the panel and then bounces back to you. Those photons do not contribute to electrical energy. Typically, the electron slopes in a solar cell or panel contain many holes in which electrons are stored. Electrons that have been drawn up to the top of the cell and that are then falling down the hill sometimes fall back into holes before they reach the bottom. Those electrons give their energy to heat or light when they fall back down the hole, and so they do not contribute to the electrical power output. In addition, the electrical holes are fairly deep. Extracting an electron from the hole requires a certain minimum energy. It turns out that one single sunlight particle or photon is required to lift out one electron, but different color photons carry different amounts of energy. The energy contained in red photons is less than the energy in blue photons, and the energy contained in infrared photons is smaller than the energy in red photons. Sunlight does not consist of one single color. Instead, it contains a mix of many colors, which can be seen if the light is sent through a prism. Many of the photons in sunlight do not have enough energy to lift an electron out of a hole. If the energy in a photon isn't enough to free an electron, then that photon will generate no electricity when it falls on a solar cell. Such photons simply heat the silicon. The quoted solar power delivery of 300 watts per square meter included all light colors, but the solar power that can really be used by photovoltaics is actually lower because not all photons from the sun carry enough energy to extract the electrons from the holes. This limitation on exploitable colors has driven research into materials than can use more of the colors available from the sun. Such efforts include studies of the effectiveness of adding quantum dots to solar panels, but such efforts have not yet resulted in commercial products. Additional research is also pursuing solar cells that separate photon-generated charges using something other than the electric field naturally created by a semiconduction junction. Some alternate systems are already commercially available. They are often cheaper than semiconductor-based systems and can have additional useful properties, like being flexible so that they can bend. Unfortunately, so far these new materials convert solar energy to electrical energy less efficiently than semiconductor-based systems.

Local Solar Generation

The previous sections argued that placing solar cells directly on the roofs of single-family homes could supply all of the average power for those homes and a significant fraction of all electrical power used in the United States; however, one cannot simply connect one's household wiring to the output of a solar panel. A system is required to convert the natural solar panel output into the form that flows out of current U.S. plugs. Photovoltaic systems function like a battery delivering a constant electric current. Such constant current devices are called direct current devices, or DC. Electrical power in the United States is delivered in the form of an alternating electrical current, that is a current that changes or alternates direction sixty times every second The voltage difference across terminals in a U.S. plug is around 120 volts. In contrast, the voltage difference across individual photodiodes are typically around 0.7 volts. Thus, powering a household plug with a solar panel requires additional electrical devices to convert 0.7 volts DC to 120 volts alternating current (AC). Because these conversion devices require electrical power to function and add to the cost of residential systems, they further reduce the total energy efficiency of a photovoltaic system. An interesting opportunity is offered by USB connectors: they supply DC power, and a new generation of USB charging is being developed to provide much higher power, when it was pointed out that such a system is more compatible with solar power than existing systems.[4]

Even if all electrical devices were magically converted to USB, most people want to use electrical energy when the sun is not shining. One option is to store energy locally. If one assumes that there are at most sixteen hours of darkness and that a household consumes an average power of 1 kilowatt, then the household must store around 60 megawatts to provide power during the darkness. If the energy is stored as gravitational energy in a water tower 10 meters above the ground, then the water tower has to hold approximately 500 tons of water. Most people find it more convenient to simply sell their unused power back to the power company, rather than storing it locally. In Germany, this process was subsidized by the government so that the rates home owners received for power delivered to the grid were greater than the rates paid by the same home owners for power from the grid. In the United States, the government did not subsidize the purchase of excess solar-generating capacity but instead offered tax breaks to offset the capital cost of the investment in the solar-powered system. Regardless of the type of incentive, some type of subsidy is probably required since a solar-powered system needs to operate for more than five to twenty years before the savings in electrical power for a household spending $150 a month on electricity exceeds the $10,000–$40,000 investment cost of the system.

Commercial Photovoltaic Systems

Large photovoltaic solar farms exist in some places, usually deserts. These systems harvest photons and convert them to electricity that is then sent

4. "Edison's revenge: The humble USB cable is part of an electrical revolution. It will make power supplies greener and cheaper." October 17, 2013, www.economist.com/news /international/21588104-humble-usb-cable-part-electrical-revolution-it-will-make-power -supplies.

Table 11. Ability of photovoltaic solar energy to provide 100 percent of U.S. energy or just electrical energy.

Quantity characteristic of photovoltaic solar power in the United States	Value	Percent of U.S. land area for 100% total power
Average power/area = 200W/m^2 insolation	200	n.a.
Average power/area = 200W/m^2 insolation × 0.2 efficiency	40	n.a.
Energy/sec/m^2 over 100% U.S. energy	3.93×10^{14}	n.a.
Fraction of U.S. area required for total U.S. energy	0.0081	0.8%
Fraction of U.S. area required for total U.S. energy excluding thermal losses	0.0041	0.4%
Fraction of U.S. area required for total U.S. electricity energy	0.0011	0.1%
No. of single-family detached houses in the United States	70,000,000	n.a.
Electric power generated if each of 70,000,000 U.S. single-family households generated 50 m^2 × 40 W/m^2 = 2 kW	$1.4E \times 10^{11}$	n.a.
Fraction of total electrical power from those rooftops alone	31.73%	n.a.
Cost/rooftop 3kW average ~10 kW peak	$20,000	n.a.
Home Depot 10 kW nameplate at $22,000 gives average $7/watt or $10,000 for average household consumption of 1,300 watts	$10000	n.a.
U.S. average electricity bill ~$100 → annual ~$1,200 so years to payoff, not including lost interest	~8 years	n.a.
Cost of 70,000,000 10 kW peak ~2 kW average rooftop systems ~$20,000	$1. \times 10^{12}$	n.a.
Cost of 70,000,000 40 kW rooftop systems ~20,000/GDP	9%	n.a.
Cost/average watt ~	$7	n.a.
Cost for 100% of total energy from photovoltaic solar	$2. \times 10^{13}$	n.a.
Cost for 100% electrical energy from photo-voltaic solar	$3.. \times 10^{12}$	n.a.
100% total energy from solar photovoltaic solar/GDP	163%	n.a.
Cost for 100% of total energy after losses 30% efficiency/GDP	82%	n.a.
100% electrical energy from photovoltaic solar/GDP	23%	n.a.

down power lines and delivered like power from a conventional electric power plant. For example, on 140 acres of unused land on Nellis Air Force Base, Nevada, 70,000 solar panels are part of a photovoltaic solar array that will generate 15 megawatts of solar power for the base. Commercial thermal solar power is discussed in Appendix B. One important feature of such systems is that they do offer the possibility of energy storage since the hot liquid can be stored and used in periods of darkness. Some solar power systems have used stored heat to generate power over a twenty-four-hour period, which is a proof concept that solar-powered generating systems can generate electricity when it is dark; however, at present, commercial thermal solar produces extremely expensive electricity, as shown in Figure 4.2.

A summary of the average power that photovoltaic solar power could provide is given in Table 11. Similar information for hydroelectric and wind power is shown in Tables 8 and 10, respectively. Table 11 shows that photovoltaic solar power could meet the entire average energy needs of the United States; however, solar power is zero when there is no sunlight, so there are significant periods during each day when solar power does not provide any energy anywhere in the continental United States. Thus, in the absence of a revolution in grid-scale electrical power storage, photovoltaic solar power cannot be the sole supplier of energy to the United States, even though it could easily meet the average energy demand.

6

COMBINING RENEWABLE ENERGY SOURCES

In Chapters 4 and 5, I argued that either wind or solar power could meet the total U.S. average energy demand; however, in the absence of large-scale energy storage, the fickle nature of wind makes it challenging to meet our instantaneous energy demands. Meeting the demands economically is even more difficult since overcapacity is the simplest way to avoid insufficient supply during periods when renewable production is low; however, such excess capacity would imply that the cost and land area estimates calculated in the previous chapters are much too low. The situation with solar energy is even worse: there is no solar power during the night.

In this chapter, I will consider how fluctuations in renewable energy supply can be reduced by pooling renewable energy harvested over large geographic regions, as well as by combining different renewable energy sources. If fluctuations are correlated, combining sources provides no improvement: the sum has the same fractional fluctuations as each of its parts. In contrast, if the fluctuations are uncorrelated, combining them will greatly reduce the fractional fluctuation, allowing the average power to equal the actual power, and allow a 100 percent renewable energy economy at an affordable cost.

In order to illustrate how the benefits of pooling sources depend on the correlations between the sources, I will consider a few simple examples based on solar power. It is possible to pool solar power from Boston and New York, but they are both in the same time zone; therefore, sunrise in New York is well correlated with sunrise in Boston. Thus, pooling solar power from both places would be fairly useless. In contrast, combining solar power from Ecuador and Malaysia would provide solar power approximately twenty-four hours a day 365 days per year since the two countries are located on the equator on opposite sides of the earth. Though such long-distance energy transfer is ridiculously impractical, the example illustrates how combining sources with different peak supply times can convert fluctuating renewable power into steady renewable power.

Customers want power on demand 365 days a year, so it is important to consider fluctuations on various timescales. Solar power has a regular daily fluctuation that governs the scheduling of most people's lives, but solar power also has strong seasonal fluctuations. In the Northern Hemisphere, days are longer in the summer than in the winter, and the reverse is true in the Southern Hemisphere. Thus, pooling solar power from both hemispheres would remove seasonal fluctuations. Combining solar power harvested at the North and South Poles would provide twenty-four-hour solar power during most of the year, even though each pole spends approximately half the year in darkness. These seasonal variations pose serious challenges for the United States and the European Union, since in both cases summer is well correlated over each region. Pooling solar power over either region does not help much. In fact, even pooling solar power over both regions would not help much.

Unlike solar power, which is always zero during the night, wind power generation is rarely zero, and systems that level power production at a rated power reduce high wind velocity fluctuation, at the expense of some energy efficiency. As a result, though wind power is not stable, a steady base supply is usually available, with fluctuations around that base determined partly by how the system is operated; however, given the power scaling with the cube of the wind speed, detailed predictions of instantaneous power delivery are extremely difficult.

Combining wind power from separate wind farms may greatly reduce wind power fluctuations on timescale of minutes and hours: a 2007 study found that the hourly correlation coefficient between U.K. wind farm sites decreases to approximately 0.1 over distances in excess of 1,000 kilometers.[1] In contrast, wind power fluctuations extending days can correlate over much of Europe, and most of Europe has low wind speeds during the summer. A 2008 study on integrating wind power over Western Europe suggested that combining European wind farms could not adequately compensate for the seasonal low in the summer.[2]

Though seasonal wind speed in the European Union has a marked lull during the summer, the seasonal wind speed variation in the United States is more diverse. For example, average wind speed in San Francisco and Reno peaks in the summer, whereas in Boston and Provincetown it peaks in the winter. In contrast, San Diego, Oklahoma City, New York City, Cleveland, Austin, and Portland, Maine, all peak in the spring, while Seattle has two almost equal peaks in the spring and the fall. Despite these local variations, Figure 4.17 showed that total wind speed averaged across the United States is lowest in the summer.

The discussion of geographical variations in average seasonal wind and solar power illustrates how combining wind or solar power from different geographical regions can reduce the variations in the supply of renewable energy since in Europe and most of the United States, wind speeds are lowest during the summer when total solar power is highest. Thus, even in a single location, harvesting both wind and solar energy may provide more stable power than either source individually since the

1. G. Sinden, "Characteristics of the UK Wind Resource: Long-Term Patterns and Relationship to Electricity Demand," *Energy Policy* 35 (2007) 112–127.

2. P. Kiss, and I. M. Jánosi: "Limitations of wind power availability over Europe: A conceptual study," *Nonlinear Processes in Geophysics* 15, (2008) 803–813, doi:10.5194/npg-15-803-2008.

two sources fluctuate in very different ways. For example, the installed solar and wind power capacity in Germany combines to remove almost all seasonal variation of each.[3]

Combining wind and solar power can reduce nonseasonal power fluctuations as well, and adding hydroelectric power to wind and solar provides an important controllable power source that can smooth out uncontrollable fluctuations remaining once wind and solar are combined.

Issues with intermittence over the course of a day are illustrated in Figure 6.1.[4] Solar power variations are fairly regular and predictable and the changes over the course of a day are small. In contrast, wind power can show large variations, which may be reduced by pooling power from distant wind farms. Figure 6.1 indicates that California is using hydropower to replace approximately one-half of solar power just after sundown.

Comparing the data for June 19 and December 14 allows comparison of winter and summer power generation in California. In total, solar generated the same amount of power on both days, whereas the wind power in June was almost ten times higher than in December. In June, the sun did indeed shine longer, but the peak solar contribution was half what it was in December, resulting in the average power being approximately the same. These results demonstrate the substantial variations in renewable power generation that pose significant challenges when using renewables to meet energy demands. As the graphs show, fossil fuel–burning plants provided most of the required adjustments, with hydropower somewhat compensating for the loss of solar PV in the evening.

Thus, mixes of wind and solar can provide more reliable stable power than either system individually. Owners of sailing ships already often

---

3. See www.ise.fraunhofer.de/en/downloads-englisch/pdf-files-englisch/news/electricity-production-from-solar-and-wind-in-germany-in-2013.pdf.

4. See www.caiso.com/green/renewableswatch.html[0]. Similar information is available for New England (http://isoexpress.iso-ne.com/reports;jsessionid=919358F9 D9DD5C08B339251B018E4762), France (www.rte-france.com/fr/developpement-durable/eco2mix/production-d-electricite-par-filiere), Spain (https://demanda.ree.es/generacion_acumulada.html), and Australia (http://windfarmperformance.info/).

employ a mixture of both wind and solar generators in order to meet fluctuating demands, where excess power is directed to desalination. Furthermore, linking larger geographical regions also lowers variation. Though the sun now sets on the British Empire, it still does not set on the world. Pooling wind power over the entire United States will lower power fluctuations associated with local variations in wind speed, and pooling consumption over the entire United States will reduce local spikes in energy demand associated with weather; however, a substantial investment in long-distance energy transport and "smart grid" optimization that matches available power with user demand would be required to make such a system effective. Combining the power generation from large-scale wind farms with the power generation from rooftop solar systems on single-family homes could provide more stable electricity generation than either source individually.

Table 12 provides quantities that are useful in evaluating the feasibility of converting the United States into a renewable energy economy. Three cases are considered: (1) 100 percent of energy currently consumed by the United States, (2) 100 percent of the energy currently consumed by the United States energy assuming that the thermal losses associated with burning fossil fuels in electric power plants and transportation do not need to be replaced, and (3) 100 percent of U.S. electrical energy. The first quantity is the maximum possible energy that would be required. The second quantity is more realistic, since approximately 50 percent of the energy content of fossil fuels is lost when they are burned in combustion engines, and combustion-based electrical turbines and the renewable-energy calculations consider energy directly used by consumers after conversion efficiency losses. Table 13 summarizes the feasibility of each wind, PV solar, or hydroelectric power meeting the current and future energy needs of the United States.

In many areas, there is a substantial push to convert to a renewable energy economy, but so far progress in the United States has been extremely mixed, with wind power showing the most rapid recent growth

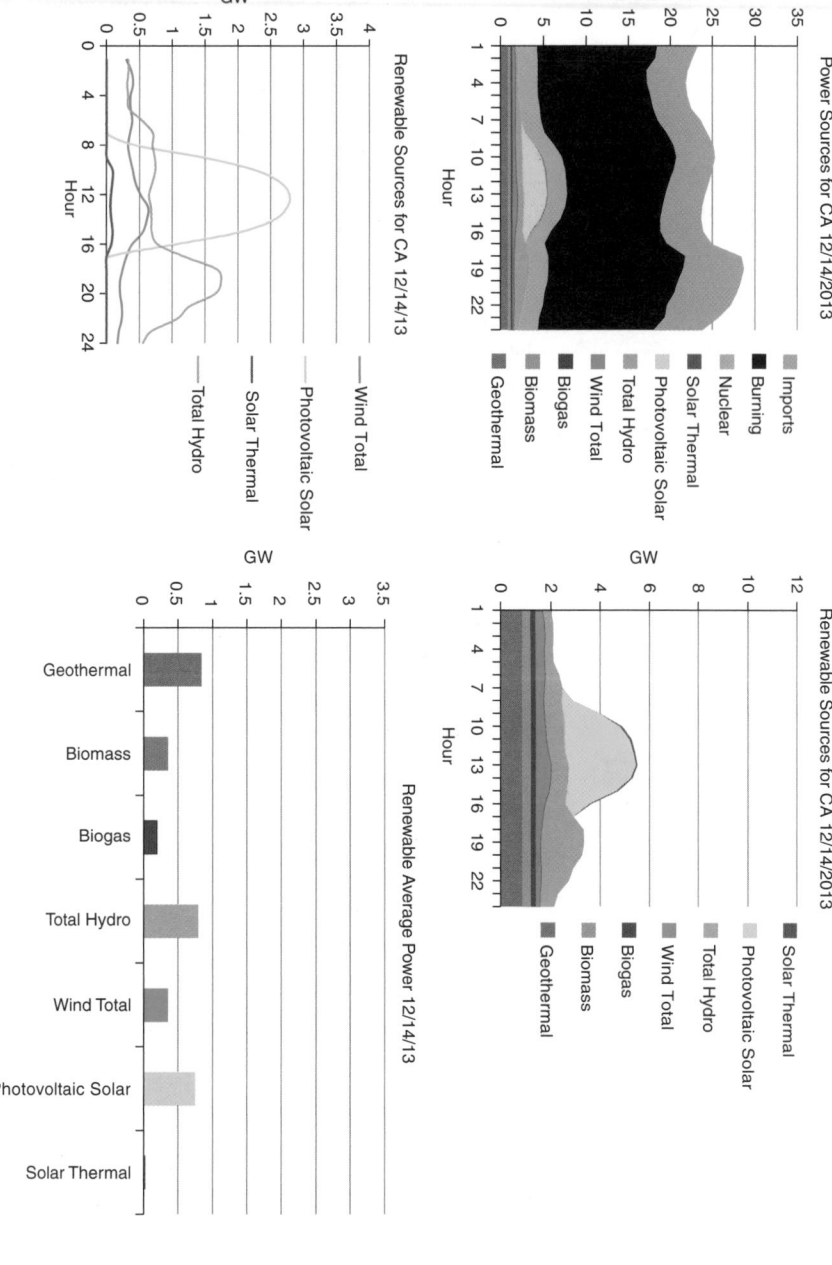

Figure 6.1 Power generation in California by hour for December 14, 2013, and June 19, 2013. The total average demand for the two days is similar, but the load in June varied more smoothly than the load in December. For

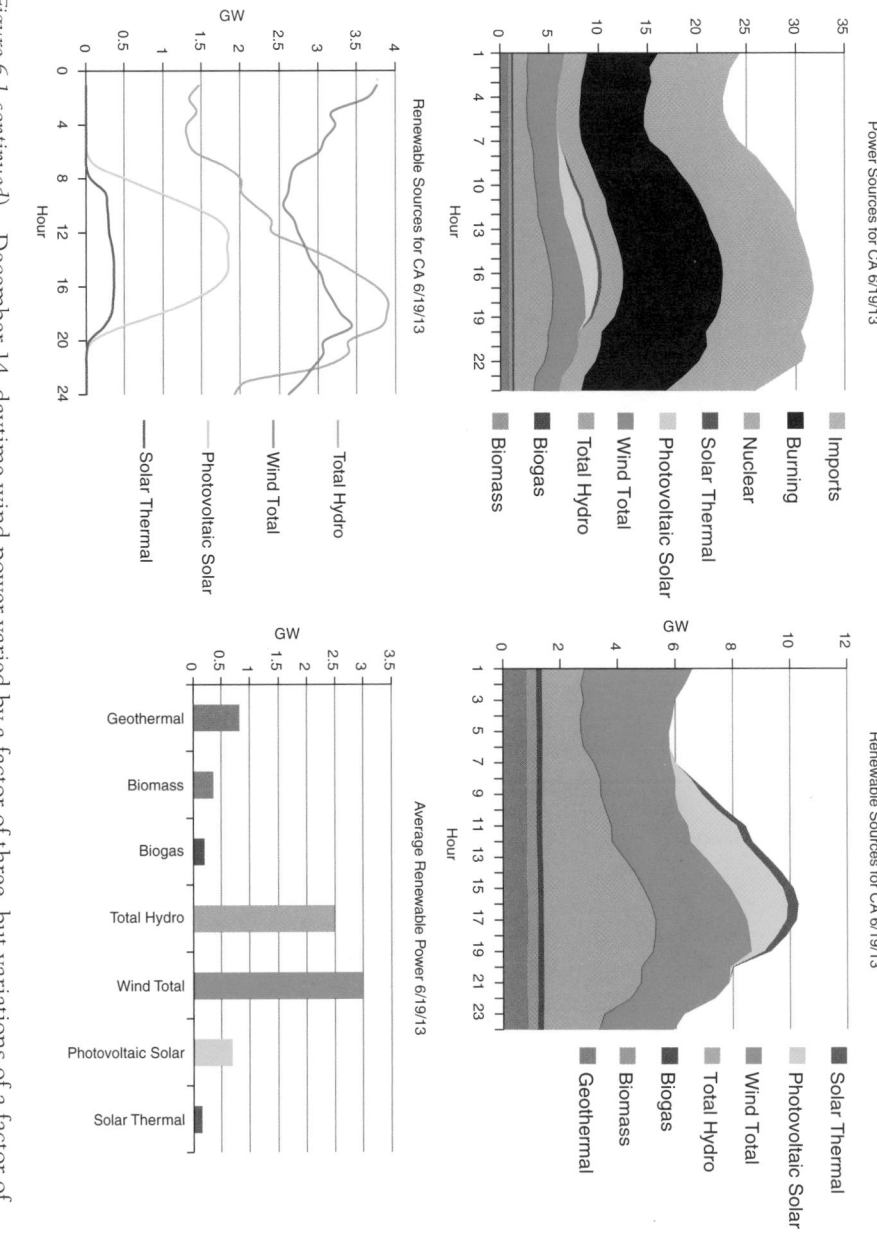

(*Figure 6.1 continued*) December 14, daytime wind power varied by a factor of three, but variations of a factor of ten are possible. Hydropower shows a large peak at night that partly compensates for the solar. The graph on the

(*Figure 6.1 continued*)   lower right shows the average contributions of each
source. The solar peak is around five times the wind power peak; however,
the average solar power contribution is only slightly over two times the wind
power contribution because solar contributes nothing during the night. For
June 19, the wind contribution is almost ten times larger than it was on
December 14, while the variation was only a factor of two, versus a factor of
four in December. In contrast, the peak solar power was only around half
what it was in December, but of course sunlight in June extends over more
hours, resulting in almost the same average solar power generation.

Table 12.   Useful quantities for evaluating the feasibility of the United States making
a complete conversion to renewable energy.

Quantity characteristic of current energy use in the United States	Value
Seconds/year	31,536,000
U.S. surface area in km^2, including water	9,833,517
U.S. surface area in m^2, including water	9.8×10^{20} E + 12
U.S. surface area km^2, excluding water	9,158,022
EIA projected total energy use for 2013 in quads	96.26
U.S. total energy consumption in joules (1 quad + 1.05×10^{18} joules)	1.0×10^{20}
Average U.S. total energy consumption in watts	3.2×10^{12}
Watts/m^2 for total U.S. energy consumption	0.34
Average U.S. total energy consumption in watts without losses	1.6×10^{12}
Watts/m^2 for energy used after 50% losses that are largely thermal	0.163
EIA projected energy required to generate electricity in 2013 in quads	38.4
EIA projected electricity consumed in 2013 in quads	12.4
EIA projected electrical energy consumed in 2013 in joules	4.0×10^{12}
U.S. electricity consumption in billion kWh	3,900
Average U.S. electricity consumption in watts	4.4×10^{12}
Watts/m^2 for total U.S. electric consumption	0.045
U.S. GDP, in billions of dollars	$15,685

during the last five years, while solar power has increased rapidly during
the last two or so years. As discussed in detail in the previous chapters,
meeting 100 percent of the total average energy needs is quite feasible
even without including efficiency reductions that result from eliminating burning, and it is very feasible if those reductions are included.

Table 13.   Potential of renewable energy based on hydro, photovoltaic solar, or wind power.

	Hydro	Wind	Photovoltaic solar
Average power in W/m^2	0.02	6.5	200
Current conversion factor	~90%	~15%–40%	~10%–60%
Conversion factor for this calculation	~90%	~30%	~20%
Average electrical power in W/m^2	0.02	2	40
Percentage of U.S. land area required for total energy	>100%	15%	0.75%
Percentage of U.S. land area required for electrical energy only	>100%	3%	0.13%
Opportunity for expansion	Not much	Yes	Yes
Land area also usable for agriculture	Yes	Yes	No
Renewable	Yes	Yes	Yes
Must emit	No	No	No
Site sensitive	Highly	Highly	Somewhat
Fundamental issues	Already almost fully used	Highly variable power	Removes all sunlight
Energy conversion limitations	None	Betz limit ~60%	Almost none
Intermittence issues	None, dams store intermittently delivered rain	Power scales as velocity cubed	1. Average daylight only 12 hours 2. Angle toward sun lowers light/area by ~1/2
Average power / peak power due to intermittent delivery	100%	30%	25%
Room for significant technological efficiency improvement	No	No	1. Increase accepted spectral range 2. Lower losses 3. Change tilt to follow the sun

Owners of existing fossil fuel power plants would not relish decommissioning such plants while they are functioning profitably, and in any case the losses involved would be passed on to consumers and taxpayers. This issue poses a challenge for the introduction of renewables since electricity use is fairly stagnant; however, a spike in electricity demand would allow renewables to provide much more energy without reducing the output of existing fossil fuel–burning plants. Converting from vehicles powered by internal combustion engines to vehicles powered by electric motors would provide just such a large increase in demand. Importantly, because of the poor energy efficiency of internal combustion engine cars, the electrical energy required for the cars would be only around 20 percent of the energy contained in the petroleum now being used to fuel cars.

In sum, meeting the average power demands is still not the same as meeting instantaneous demands given the lack of practical energy storage. Even if the entire continental United States is linked as a single energy grid that optimally combines wind power, solar power, hydropower, and geothermal power to meet instantaneous demand, in the near future more economical to burn some natural gas than to convert 100 percent of our power sources to renewable energy.

7

DISTRIBUTING ELECTRICITY

To paraphrase a mid-twentieth-century detective novelist, our romantic lives and our electrical delivery systems are similar: if they are working well we ignore them, but if they are not working well we can think of little else. In industrialized countries, electricity is distributed to almost all residences and businesses using wires. This conveniently delivered energy allows us to heat, cool, and light our houses. It also runs our appliances and powers our electronic devices. Many people hope that in the future it will provide the energy that will run our cars. Yet, for all of the importance of electrical power, we rarely pay much attention to it.

As discussed previously, commercial electrical power is provided by a current of electrons flowing along wires. In SI units, that flow of electrical current is measured in amperes. The electrical power consumed by a device is given by the product of the current, I, and V, the voltage decrease resulting from flowing through the device. In equation form this becomes

$$\text{Power} = I \times V.$$

The voltage difference corresponds to the difference in electrical energy between the two sides of the device. Similarly, if g is the gravitational constant, then $g \times h$ corresponds to the gravitational potential energy difference per unit mass between two positions with a height difference of h.

It turns out that "voltage" has no absolute meaning. The only thing that is meaningful is the difference between two voltages. A similar issue arises in connection with gravity. We can call sea level height zero, in which case Denver has a height +1 mile, or we can call Denver height zero in which case sea level has a height of −1 mile. In both cases, Denver is one mile higher than sea level. If one is trying to lift a weight from sea level to Denver, it does not matter if one describes the process as moving from −1 mile to zero or from zero to +1 mile because the same amount of work is required. A similar situation occurs for voltage. People often speak of 1 volt or 5 volts or 100 volts, but those numbers are meaningful only if they are compared with something. Just as most people define sea level as zero height, in which case the Dead Sea and Death Valley are at negative heights, most people define "zero voltage" as the electrical potential energy of the earth. Sometimes people who are careful actually refer to this as "earth ground." In every house, there is a connection to a pipe (usually a copper water pipe) that is in electrical contact with the earth. This represents the "earth ground" value for that house. The voltage delivered to a house is described as the potential difference between the wire from the power company and earth ground. Since all of the houses are connected to the same earth, this earth ground value is universal. Of course, the power company is also attached to earth ground, so most discussion of electric power simply refers to a voltage, even though what is really meant is the voltage with respect to earth ground.

For most electricity delivery systems in the world, the voltage delivered can be expressed as

$$\text{Voltage Out} = V_0 \sin(2\pi t/\text{Period} + \phi),$$

where the voltage out is the potential difference between earth ground and the wire. In the United States, the period is one-sixtieth of a second, so the system completes sixty cycles in one second. In Europe, the period is one-fiftieth of a second, so the system completes fifty cycles in a second. Such systems are called alternating current or AC systems: the direction of the current alternates with time because the voltage oscillates between positive values, which pull electrons from the ground toward the power plant, and negative values, which push electrons away from the power plant toward the ground.

INSIDE THE HOUSE

In order to use delivered electricity, people plug devices into wall plugs. Figure 7.1 is a picture of a typical grounded U.S. wall plug. Each outlet has three holes: two long vertical holes and one hole that looks like a D tipped on its side. The D is the earth ground. The little slot is the "hot," which means that its potential difference with respect to the ground is given by the equation above. The longer slot is called neutral. The size difference between the neutral and "hot" slots makes it impossible to connect an electrical device backward with the hot connected to the neutral. This is a safety feature to avoid electrical accidents.

When something is plugged into a wall socket, the electrons flow from the "hot" and move through the electrical device. The electrons that enter a device from the "hot" wire leave the electrical device through the neutral wire.

For the United States, the voltage difference between the hot and ground in a wall plug is described by the following equation:

$$\text{Voltage} = 170 \sin (2\pi\ 60t + \phi),$$

which is indicated by the red line in Figure 7.1. For convenience in electrical calculations, the voltages supplied by the power company are usually described in terms of their root-mean-square value, which is their peak value divided by the square root of 2. For the United States, this value is 120 volts, which is why most people think of the United States as a 120-volt system. In contrast, Europe uses a 240-volt system, which means

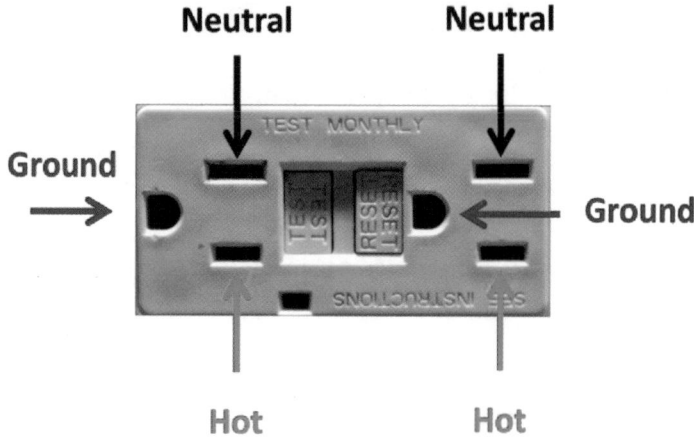

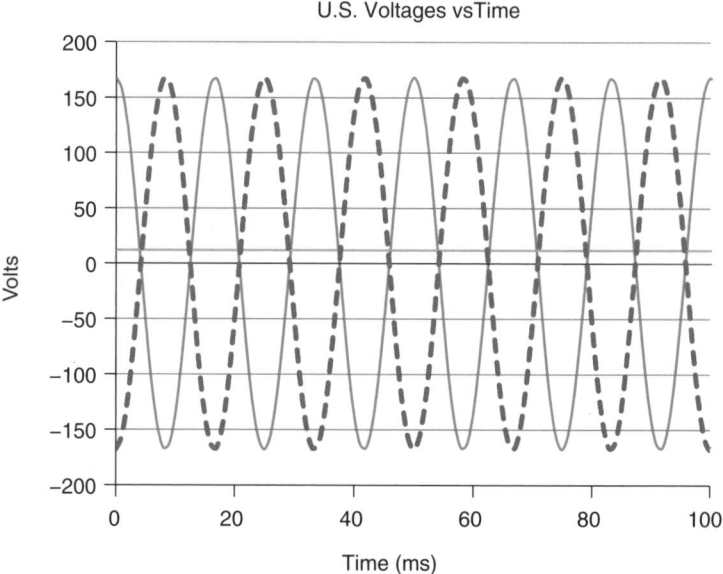

Figure 7.1 Illustration of electric power delivery to U.S. consumers. The upper image shows a standard U.S. grounded plug, highlighting the connections for the ground, neutral, and hot wires. The lower image shows voltage differences as a function of time. The red line shows the difference between the hot and the neutral, and the black line shows the difference between ground and neutral. The horizontal gray line shows the output of a typical solar cell, which does not change with time. The approximately 17-volt constant solar panel output appears as a straight line just above zero volts. Clearly, converting solar panel outputs into wall plug outputs requires additional processing, which is part of the cost of a solar panel system.

a peak-to-peak excursion of more than 600 volts. This high European voltage used to pose serious problems for Americans traveling in Europe because electrically powered equipment designed for one place would often not work correctly in the other. Worse, the systems could sometimes fail catastrophically, destroying the device and potentially injuring the user. Now most electronic devices use digital circuitry to manage input voltages, so computer power supplies and cell phone chargers can be plugged in either in Europe or in the United States without difficulty. All that is required is a different plug; however, it is useful to know that when plugging a U.S. device into a European circuit the U.S. plug should be inserted into the converter before the converter is inserted in the European plug. The reverse order frequently results in rather unsettling spark generation.

In contrast with the time-varying voltages characteristic of the generating systems that exploit Faraday's law, the solar panel output varies with the amount of sunshine hitting the panel, over the less than one-second time scale shown in Figure 7.1. The output of a solar panel is effectively constant over minutes, though it changes slowly over a day as the sun rises and sets. Such constant current sources are called direct current, or DC sources. A typical voltage from one individual cell is approximately 0.5 volts; however, in commercial solar panels individual cells are combined to provide somewhat higher voltages. Outputs of 12 to 24 volts are not uncommon. The gray blue line in Figure 7.1 shows a 17-volt constant output versus time for a typical solar panel. Obviously, this output voltage is not the same as that provided by wall plugs. Thus, if one wants to replace the power provided by U.S. power plants with the power generated from a solar panel on a building's roof, additional electronics will be required to convert around 12 volts of fairly time-independent voltage into a time-varying voltage that reaches 170 volts sixty times a second.

Wind sources use rotating shafts and Faraday's law to create voltage that varies in time as the shaft rotates; therefore, like fossil-fuel-burning electrical power stations and hydroelectric power plants, wind turbines naturally produce a time-varying voltage. In wind-driven systems, the faster the wind is blowing, the faster the generator shaft rotates, increasing the number of voltage cycles per second. Of course, one can use gear boxes and other techniques to reduce the variation with wind speed, but if wind power is to be supplied to the U.S. power grid and delivered to customers over the same wires as power generated by fossil fuel–burning plants, then the voltage generated by the wind turbines has to have the same time dependence as the voltage from the fossil fuel plants. In practice, it is often easier to convert the wind-generated voltage from a time-varying or alternating current (AC) signal to a constant-voltage or direct current (DC) signal. The power can then be shipped as DC and converted back to AC to match the fossil fuel plants that contribute power to customers. A later section of this chapter will consider different long-distance electricity-delivery strategies, but the next section will consider the last interface between the power company and an individual consumer.

JUST BEFORE THE HOUSE

In the United States, power is transported to customers at much higher voltages than the 120 volts that are characteristic of wall plugs inside houses. As will be discussed in the next section, high voltage transport is used because the losses in the wires decrease as the voltage increases. At present, in the United States, the delivery voltages are usually 7,200 or 14,400 volts, but there is a push to increase those voltages in order to increase energy efficiency. In order to be useful to consumers, the high voltage that passes through the wires from the power plant must be converted into a 120-volt source for U.S. wall plugs. The devices that make this voltage change are called transformers.

Transformers use Faraday's law, which says that a time-changing magnetic field flux through the area enclosed by a wire loop generates an electrical current that will flow around the wire loop in which the magnetic field flux is changing.[1] The time-changing magnetic flux in the secondary loop causes current to flow in the secondary loop, as illustrated below:

Time-changing current from electrical company
→ time-changing magnetic field in the space occupied by
the secondary loop
→ time-changing magnetic field flux through secondary loop
→ current flows in the secondary loop in accordance
with Faraday's law.

The two wire loops are not electrically connected. Faraday's law allows power to be transmitted between the coils without the coils touching. Since Faraday's law depends on a time-changing field, such a power transfer system works only in AC systems. This is the wonder of AC power systems. It is possible to take electrical power from a wire without ever making electrical contact with the wire. Shorting out the secondary loop does not short out the primary loop. This is extremely important. It means that it is easy to connect and disconnect people from the power grid since the wires that are connected to the power plant are never touched. DC systems must use a different strategy.

One can control the ratio of the voltage in to the voltage out by controlling the number of loops in each wire. The voltage out is equal to the voltage in times the ratio of the number of loops in the secondary loop divided by the number of loops in the primary. If the secondary has fewer

---

1. In electrical generators, the time-changing B field flux was created by rotating a wire loop with respect to a stationary magnet. In a transformer, there are no moving parts. The current flowing in the primary coil generates a magnetic field, just as the current flowing in the wire loops in electric motors generates a magnetic field. The magnetic field is proportional to the current flowing in the wire loop, so when the time-changing AC current is applied to the wire loop, it generates a magnetic field. That time-changing magnetic field flows into a secondary wire loop. The time-changing magnetic field created by the current flowing in the primary loop creates a time-changing magnetic field flux in the secondary loop.

loops, the output voltage is smaller. If it has more loops, the voltage is higher.

If the device were perfectly efficient, the power in would be equal to the power out. The electrical power delivered to a device is characterized by the product of the current times the voltage as shown in the following equation:

$$Power = Current \times Voltage = I \times V.$$

Because electric power is the product of the current times the voltage, in a system with no losses the product of the current and voltage in would equal the product of the current and voltage out.

A photograph of a pole transformer along with a schematic of its internal structure are shown in Figure 7.2. The schematic in Figure 7.2 shows that electrical power transformers consist of two or more coils of wire that are not electrically connected to each other. The power from the power company passes through the primary coil, and the houses are connected to two secondary coils. The lack of connection is very important since it prevents a short circuit in the house from shorting out the entire power grid. This physical separation is a very important feature of AC power.

In Figure 7.2, the primary coil is connected to a 7,200-volt source and the secondary coil provides the ±120-volt electricity that connects to breaker boxes in houses. Thus, the output of the transformer is two 120-volt signals that are 180 degrees out of phase with each other as illustrated in Figure 7.1, where the red line corresponds to the red transformer output and the purpled dashed line corresponds to the purple transformer output, which is 180 degrees out of phase with the red output.[2]

---

2. The 240-volt differences required for large appliances like electric dryers are provided by the voltage difference between the purple voltage and the red voltage in Figure 7.1.

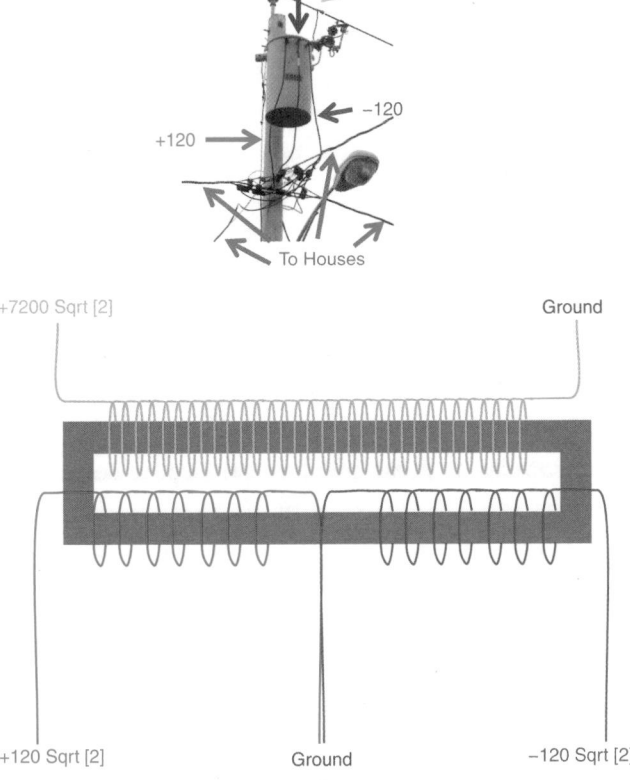

Figure 7.2 Photograph of a U.S. power transformer and an illustration of the inside of the transformer where the 7,200 input voltage to the transformer is converted into the 120-volt output voltage that is delivered to wall plugs inside the house. The high voltage comes into the transformer from the power station or from a substation. The transformer sends three wires to each house. One wire connects the grounds. The other two wires deliver power to the houses. The lower panel shows a schematic of the inside of the power transformer. The transformer functions following Faraday's law: the time-changing voltage from the power station passes through the wire coil shown in orange. The resulting current creates a time-changing magnetic field that results in a voltage across the red and purple wire coils. The three wires going to each house correspond to the three wires leaving the bottom of the figure. Notice that there is no physical connection between the orange coil and the red or purple coils. This lack of physical connection is an important feature of AC power delivery systems.

LONG-DISTANCE POWER TRANSMISSION

Fossil fuel–generated electricity is usually not transported far because the energy loss due to transmission increases with distance, and fossil fuel plant performance is largely insensitive to location because fuel burns anywhere. In contrast, hydroelectric dams need to be located in suitable locations, which are often quite distant from population centers. It is not enough to just generate the power. It is necessary to deliver the power to consumers who will pay to use it. A similar issue arises with wind power because wind power scales with the cube of the wind speed. Furthermore, linking sources and customers over the entire continental United States could make the conversion to a 100 percent renewable energy economy much more feasible by smoothing out local variations in energy supply and demand. All of this linking requires physical connections using metal wires.

HIGH-VOLTAGE TRANSMISSION REDUCES LOSSES

The wires that deliver the power have a finite electrical resistance. That means that electrons traveling down the wires collide with the atoms in the wires and transfer energy to them, just as electrons transfer energy to filaments in incandescent lightbulbs or heating elements in electric heaters and stoves. If the wire has a resistance R ohms, and the current flowing through the device is I, then the power lost to the wire is given by

$$\text{Energy Loss} = I^2 \times R.$$

Obviously, lowering the resistance of the wires lowers this loss. Superconducting wires have no resistance, but at present they are extraordinarily expensive and quite impractical for electrical transmission; however, the possibility of lower electrical losses drives research on superconducting wires. If we are stuck with R, the only way to lower the loss is to reduce the current. The current can be reduced by consumers using less power;

however, it is useful to consider the case where the consumer's demand for power is unchanged. In what follows, I will show that this energy waste can be reduced without lowering the power delivered to the consumer, if the electricity is transported along wires at higher voltages.

In order to demonstrate why higher voltages result in lower power waste, it is important to start with the power required by a consumer and then determine how that power could be delivered with the smallest possible loss. For specificity, consider the power required to run a 1.2-kilowatt microwave. Since power is equal to the product of current and voltage (e.g., Power $= I \times V$), if the voltage coming out of a U.S. wall plug is 120 volts, then the current required to power the microwave is

$$I = \text{Power}/V = 1{,}200 \text{ W}/120 \text{ V} = 10 \text{ amps.}$$

If that power traveled down a wire with a total resistance of 1 ohm, then the energy wasted in the wire is given by

$$\text{Energy Loss} = I^2 \times R = (10)^2 \times 1 = 100 \text{ W.}$$

In contrast, consider the loss if the power is transported along the wire at 120,000 volts. To deliver 1.2 kilowatts at 120,000 volts requires only 0.01 amps. Thus, the power lost in the 1-ohm wire is only as follows:

$$\text{Energy Loss} = I^2 \times R = (0.01)^2 \times 1 = 0.0001 \text{ W.}$$

Thus, increasing the voltage by a factor of 1,000 reduces the energy waste by a factor of 1,000,000. Of course, a transformer is required to convert the transmission voltage to the delivery voltage, but the loss due to the transformer is usually much smaller than the loss reduction due to high-voltage transport along the wires.

Given the value of transporting electricity at high voltages, one might wonder what the upper limit on that voltage is. At voltages larger than 2,000 kilovolts (i.e., 2,000,000 volts), corona discharge losses are so large that they can offset the lower resistance loss in the line conductors. In the United States, power line voltages can range from 100,000 volts to up to

765,000 volts for long-distance transmission. As a result, transmission losses are small. According to the EIA, the average annual transmission and distribution loss in the United States is around 7 percent, and an additional 5 percent of the power is lost to provide power for the generating plant itself.

At the beginning of the twentieth century, there was an enormous controversy over whether electricity should be delivered as an AC voltage or a DC voltage. The lightbulbs that Thomas Edison invented work in both systems, but Edison argued passionately in favor of DC systems, including vivid discussions of the likelihood of death by electrocution in systems using AC power. Our present electrical distribution system at the consumer level is dominated by AC; however, DC is making a comeback in two areas: final consumer consumption and long-distance transport. The ubiquitous devices that receive their power from USB power cords are using DC power. Thus, for end users, DC power is becoming ever more important. If rooftop solar becomes common, it may be advantageous to directly feed DC power from solar panels into DC-powered devices without first converting into AC to provide power to the wall plug and then converting back into DC to provide power for the device. Each energy transformation that is avoided reduces losses.

Not only is local DC power generation and consumption increasing, but long-distance DC transport is increasing as well even though the electricity is normally generated and consumed in the AC form,[3] so conversion to DC is required before long-distance DC transport can occur. Some energy losses are inevitably associated with this conversion, so the conversion is worthwhile only if the energy loss due to long-distance transport is reduced by more than the conversion loss.

One way to understand when DC power delivery is favorable to AC delivery is to consider the analogy between electrical power delivery to

---

3. Photovoltaic solar is the one DC generation system. All of the others use Faraday's law, which requires that the output current changes with time.

shipping of material goods. In the last fifty years, container shipping has become dominant. In container shipping, enormous ships deliver cargo packed in containers. This system is optimized for long-distance hauling, from China to the United States, for example. Such ships do not stop off at every port and deliver a few iPhone chargers. That would be enormously expensive and inefficient. Rather, the big ships travel from very large port to very large port. In Europe, there are only about five ports that can receive them. The goods that are shipped are manufactured in various places and then shipped to the large port of embarkation via truck, train, or a combination of train and truck. After the goods are loaded on the ships, they are transported together by sea. Once the goods arrive at their destination port, the individual shipping containers are offloaded and transported by truck, train, or train and truck to a warehouse. From there, the individual parts are sent out to retailers, who in turn deliver them to consumers, who transport them to their homes. This sort of container shipping is a big advantage if lots of goods are being transmitted together over long distances because shipping by sea is the least expensive form of long-distance transport; however, if a pizza parlor is delivering to a customer a block away in New York City, it makes no sense to gather all of the pizza going north in the city and put it on a barge that moves it north along the Hudson River and then offload the pizzas from the barge and deliver them. It is much more sensible to simply have a bicycle messenger deliver the single pizza, even if bicycle delivery is less energy efficient.

Similarly, the big advantage of AC power is that it is very easy to take off a little bit of power at the location of each consumer without affecting the grid because it is possible to remove power from an AC power line without making a physical connection to the wire. Unfortunately, the same effect makes AC power lines constantly lose power as the energy is transported because the power radiates from the wire to nearby conducting surfaces, including the earth. Most people are familiar with the energy loss due to electromagnetic radiation in the context of cell phones: cell phone batteries die faster if one spends lots of time talking on the phone because sending out a cell phone signal requires energy. This energy loss is a direction consequence of Maxwell's equations, which state that time-changing electro-magnetic fields radiate energy.

In contrast, DC power lines do not radiate any power because the fields in them do not change with time. Thus, DC power transmission along wires is more efficient than AC transmission. DC power lines can also be located lower to the ground and take up less space than AC power lines. The power loss problem in AC transmission lines is even worse in undersea cables because salt water is an even better conductor than dirt and the salt water is located right next to the undersea cable. However, since almost all electrical consumers require AC power, the DC power transmitted over long cables still must be converted to AC at the end. Furthermore, with the significant exception of solar, most powers sources are AC. Thus, to transmit DC power, one must convert the AC power to DC power, transmit the power, then convert it back to AC at the end. This is advantageous if the power is transmitted such a long distance that the increased losses due to transmitting AC power dominate over the energy losses inherent in converting AC to DC and back. Thus, DC power transmission is analogous to shipping containers in a large container ship: it is economically advantageous when lots of power is transmitted a long distance, but it is not advantageous in the delivery of a few objects over short distances. In practical application, power from many windmills or solar panels is gathered together at the source and fed to one large DC power line that transfers it to a power station near a population center. Once at the power station, the power is converted back from DC to AC and added to the power in the local grid, which then delivers the power to individual consumers. In contrast, for local energy transport, it is more energy efficient to deliver AC power generated at a local power plant directly to the consumer as AC power, rather than to convert the power to DC, transport it, and convert it back again.

There are four big factors contributing to a resurgence in DC power: (1) the strong siting requirements for many alternate energy sources, which means that electrical generation states must be located far from the population centers that are eager for the electricity; (2) the increased cost of energy and the increasing importance of energy conservation, which has made creation of additional transmission infrastructure economically desirable; (3) the improved availability of systems that convert between DC and AC as well as between AC and DC and even from DC to DC at a different voltage; and (4) the need to combine power from different AC systems whose time dependence might not be synchronized.

Combining electricity from many separate power sources using a grid works very well as a power distribution system because it allows sharing and provides redundancy. If a power company needs to take a power plant or a transmission tower off-line for maintenance, the other parts of the grid can pick up the slack. Combining power from renewable energy sources distributed over wide areas also reduces fluctuations in power generation. For example, the correlation between wind speed begins to decrease when turbine separations exceed around 600 miles. Furthermore, if solar panels were located all around the earth, solar power would be available 100 percent of the time. Thus, the ability to maintain grids that combine power generated over very large areas may make the transition to renewable energy more feasible.

An important feature of the power grid is that it cannot store any power anywhere in the system. At any moment, you have millions of customers consuming megawatts of power produced by dozens of power plants whose outputs are coordinated so that in aggregate they produce exactly the right amount of power to satisfy that demand. The system is redundant, so failure of one piece does not generally change the power delivered to consumers; however, catastrophic failures that can black out major urban areas, like New York City, for days can occur. Redundancies add expense, but they add robustness. Achieving the correct tradeoff between the two depends on what cost one places on failure to deliver power for a given period of time. As computer power increases and power grids combine sources from larger areas, smart grids may provide significant efficiencies by allowing plants to operate at maximum capacity where they are most efficient.

8

CONSERVING ENERGY

Lighting is an area where a technological revolution has made enormous efficiency improvements possible. Governments in the developed world are aware of this opportunity and have been enacting legislation to eliminate incandescent light bulbs.[1] However, it is important to note that even if lighting consumed no electricity, residential electricity use would be lowered by only 14 percent. Thus, identifying and developing ways to improve efficiencies in the performance of other sources of electrical energy consumption beyond just lighting would be required to reduce energy consumption by a larger factor.

LIGHTING

At present, lighting is undergoing a technical revolution. In the days before artificial lighting, human activity substantially declined after dark. Candles and oil burning lamps allowed people to do more after dark, but they were hazardous and could be substantial sources of indoor air pollution.

1. See www.articles.latimes.com/2014/jan/01/business/la-fi-mo-incandescent-lightbulb-ban-20140101; www.usatoday30.usatoday.com/tech/news/2007-02-20-australia-bulb-ban_x.htm?csp=27; www.theglobeandmail.com/report-on-business/industry-news/energy-and-resources/canadians-stock-up-on-incandescent-bulbs-as-ban-takes-effect/article16293750/.

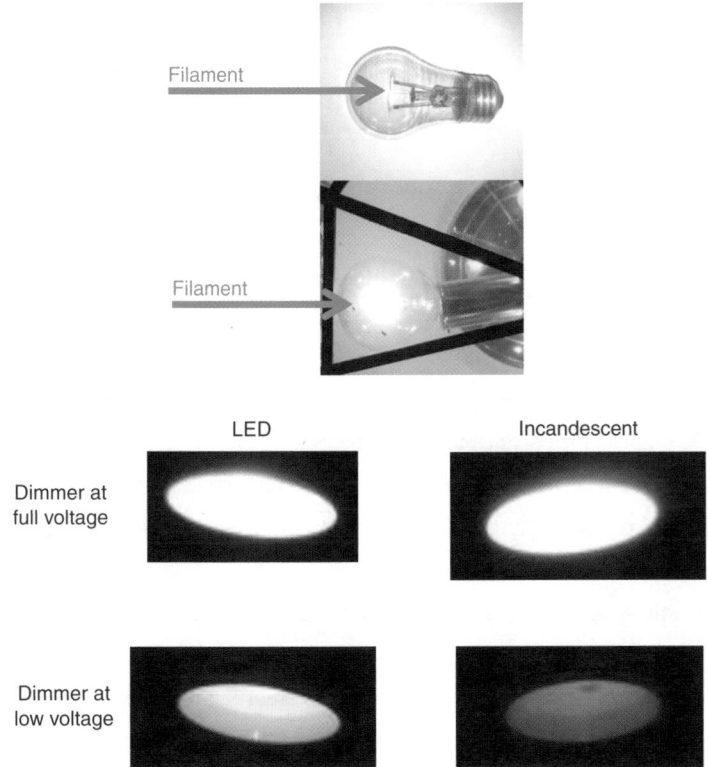

Figure 8.1 Illustrations of incandescent lightbulbs. The first image is an actual photograph of a bulb, showing the filament. The second image is a picture of a lit bulb, where the light is clearly emerging from the glowing filament. The lower box shows two photographs of the light emitted by an LED and an incandescent bulb on the same dimmer. In the upper images, the voltage was turned fully up to 120 volts. At that voltage, both the LED and the incandescent bulb emit fairly white light; however, at the lower voltage the incandescent light turns a deep red, whereas the LED remains almost white with a slight yellow-green tinge.

The development of the incandescent lightbulb made cities "that never sleep" possible. Figure 8.1 illustrates the basic features of incandescent lightbulbs. They consist of a glass bulb that has been evacuated so it contains almost no air molecules. Inside the bulb is a thin wire filament. The filament is composed of a metal that has a high electrical resistance, which means that electrons passing through the material experience a lot of friction because as the electrons move through the metal, they frequently collide with the atoms in the metal. Each collision transfers some energy from the moving electrons to the metal, which makes the metal ions move. The same effect also produces heat in electric heaters and stoves.

A similar effect is seen when a boxer hits a punching bag. The collision between the hand and the bag causes the bag to vibrate back and forth. Hitting harder transfers more energy to the bag, which makes it bounce back and forth more energetically. In the lightbulb, that energetic vibration of the molecules means that the temperature of the metal has increased. This sort of frictional heating is referred to as joule heating, in honor of James Prescott Joule, who demonstrated that electrical energy could be transferred to heat.

Filaments glow simply because they are hot. Any hot object will glow with a color that is determined only by the temperature. Most people have observed this in connection with electrical stoves or heaters: at low power, they appear red, but as the power is cranked up, they become orange. Electric heaters sometimes can even become yellow. Objects whose emission is characteristic of their temperature are called "black bodies." Astronomers often determine the temperature of stars based on their color. Light from incandescent lightbulbs reminds us of light from the sun because the temperature of the hot filament is similar to the temperature at the surface of the sun.

The relationship between color and temperature is illustrated in the lower part of Figure 8.1. When the dimmer setting is low, the incandescent light is much redder than it is when the dimmer is turned up so that the full voltage reaches the bulb. Humans like incandescent light because at full voltage the colors are similar to those in sunlight. Dimmers lower the amount of emitted light because the lower-applied voltage does not heat the filament as much as the full voltage. As a consequence of the lower temperature, the resulting color is red rather than white.

A summary of the energy transfer process for an incandescent light is shown below:

Electron kinetic energy → motion of the atoms in the
metal filament → visible photon emission

Incandescent lightbulbs are incredibly wasteful because generating white light requires heating the filament inside the bulb to approximately the temperature at the surface of the sun. Energy efficiency can be measured in terms of lumens, a measure of how much light is emitted, per watt. For a typical incandescent lightbulb, the figure is between 10 and 20 lumens per watt because so much energy is lost to heat.

Fluorescent lights and LED lights, because they are hot, do not emit light and are therefore more efficient. Instead, they transfer electrical energy to light energy using more complex strategies that are still much more energy efficient. Both fluorescent lightbulbs and incandescent lightbulbs feature glass enclosures; however, in incandescent bulbs the inside of the enclosure is evacuated, whereas fluorescent lightbulbs are filled with atomic gas, which is usually mercury. An initial large voltage creates a plasma inside the bulb. The electric power provided to the bulb accelerates particles, which collide with the gas atoms, which in turn become excited. The atoms then deexcite by emitting light at a frequency characteristic of the atoms. Similarly, if one excites a guitar string, the frequency of the sound that the guitar will emit is characteristic of that string. In other words, a D string emits a sound characteristic of D. For mercury, the excited atoms emit ultraviolet photons at two particular frequencies. We can't see ultraviolet light, so the light emitted by the mercury must be converted to visible light. This is done by the atoms in the white coating inside fluorescent bulbs. Those atoms absorb the ultraviolet light, which placed their electrons in an excited state. Those atoms then spontaneously emit light at a few particular visible colors. A list of the six steps in the energy transfer process in fluorescent light is shown here:

1. Electrical potential energy (voltage)
2. Kinetic energy of charge particles
3. Excited electrons in mercury atoms
4. Electrons return to their ground state by emitting an ultraviolet photon
5. Ultraviolet photon is absorbed by a phosphor, atom creating an excited electron
6. Electrons return to their ground state by emitting a visible photon

For fluorescent lights, dimmers do not change the color of emission because the emission colors are determined by the energy structure of the atoms in the lightbulb, which are not affected by the voltage. Fluorescent lights can generate 30 to 100 lumens per watt, with compact fluorescents generating approximately 50 lumens per watt. Thus, on average, they are more than twice as efficient as incandescent lights; however, they do not show the nice warm broad white light characteristic of the sun. As a result, people dislike fluorescent lighting.

LED lights represent a technological revolution. Like fluorescent lights, LED lights are not black body emitters. For compact fluorescents, the emitted colors depend on the type of atoms used in the bulb gas and coating. For LEDs, the emission depends on the crystal structure of the material emitting the light. Originally, LEDs used natural crystal structure, which determined the colors that they could emit; however, developments in nanotechnology now allow the fabrication of artificial crystals whose structure is controlled by the fabricators. This has allowed the creation of LEDs that emit at a variety of different colors.

Just as motors and generators can be the same object depending on whether you are providing mechanical work to turn the shaft and generate electrical energy or using electrical energy to turn the shaft, LEDs and solar panels can be the same object, though like motors and generators they are usually optimized for one application or the other. When they function as a solar cell, sunlight does work to create free particles in a region that normally has no free charge. That work is required to pull the electron out of a hole where it is trapped. Once the electron is free from the hole, it falls down the electrical hill, acquiring energy that allows the electrons to do work, as illustrated in the right half of Figure 8.2. In contrast, when the device is functioning as an LED, a voltage does work to push the electron up the hill through the depletion zone, where there are no free holes into which it can fall. After it gets to the top of the hill, it is at the end of the depletion zone. The electron then encounters a mesa full of holes, which is the p doped region of the semiconductor, and it rapidly falls into one of the holes. As it falls, it gives up its energy by emitting a photon.

The LED emission process is spectacularly efficient. System efficiencies can exceed 90 percent for single-color LEDs; however, just as fluorescent lightbulbs emit only a few colors that are determined by the atoms in the lightbulb, LEDs emit light whose color is determined by how deep the holes are. As a result, an LED naturally emits only one color. In order to make the LED emission colors more palatable, various energy efficiency losing strategies are used. As a result, for U.S. residential lighting applications, LED efficiencies are comparable to fluorescent efficiencies with 60 to 100 lumens per watt. "Cool lights," which appear blue tinged, give 60 to 90 lumens per watt, whereas the preferred "warm lights" give 27 to 54 lumens per watt.[2]

LED lighting has revolutionized lighting in countries in the developing world. A solar panel coupled with an LED can light the darkness for twenty years without creating indoor air pollution or presenting a fire hazard. Such systems can pay for themselves in approximately a year.

2. See www.energysavers.gov/your_home/lighting_daylighting/index.cfm /mytopic=12030.

LED Analogy　　　　　　　Solar-Cell Analogy

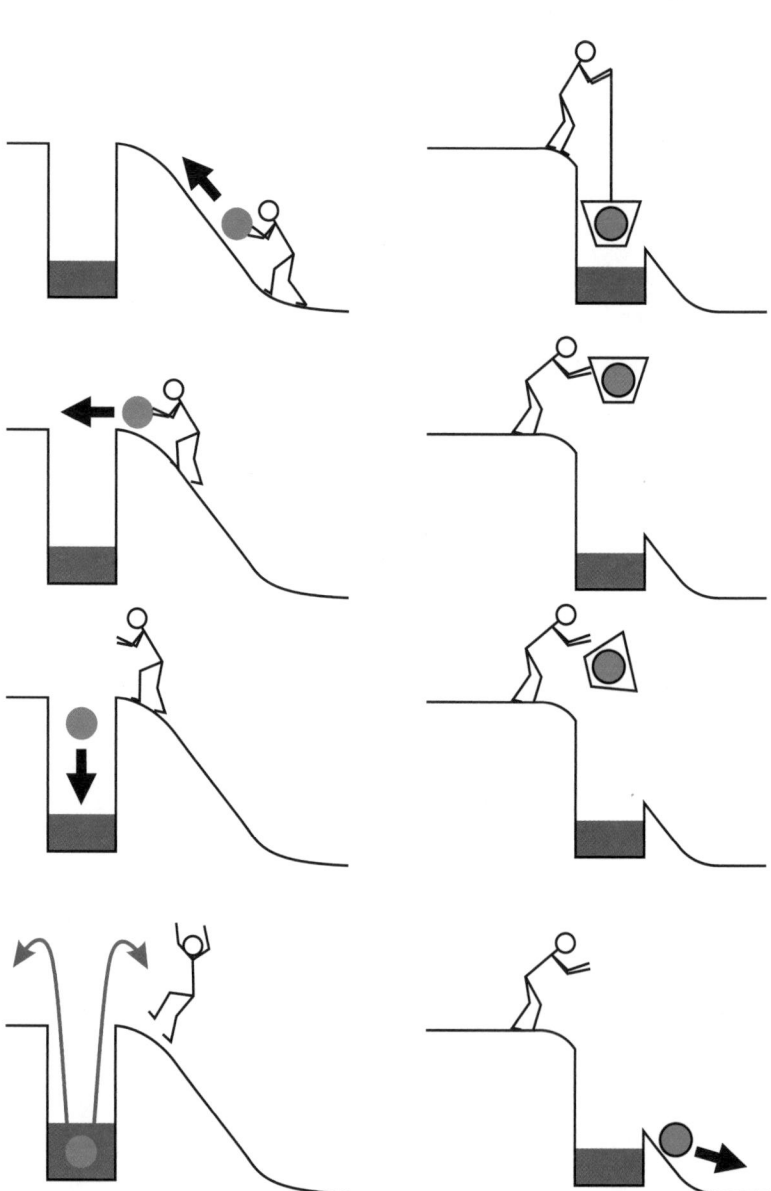

Figure 8.2 Schematic of how LEDs and solar cells are exactly the same devices. In the solar cell case, light does work to free charges that are trapped in holes. Once the holes are free, they gain energy by moving down an electrical hill. In LEDs, the voltage source pushes charges up the hill. At the top of the hill, they fall into holes, emitting light as they fall.

INTERNAL COMBUSTION
BURNING TRANSPORTATION

Like the Carnot engine, internal combustion engines use the pressure generated by the expansion of heated gas to press against a piston, which then does mechanical work. In the Carnot engine, the working fluid of the engine, frequently water, is physically separated from the burning fuel. In contrast, in internal combustion engines, fuel burns in a chamber. The interaction between the fuel and the oxygen in the air heats the air inside the chamber. The pressure that results from the hot air can then do mechanical work to push a piston or a turbine. One of the reasons that gasoline has such a high energy per volume is because the oxygen required for the burning chemical reaction is supplied by the air. In contrast, batteries must carry all of the required chemicals inside a package that also has mass.

Cars

Internal combustion engines are cyclic heat engines, so they are governed by the Carnot efficiency limit. Like the turbines that drive electric power generation, internal combustion use heat energy to expand a gas, where that gas expansion performs mechanical work. Most cars use a four-stroke engine, with four basic steps that repeat with every two revolutions of the engine: (1) intake stroke, (2) compression stroke, (3) power stroke, and (4) exhaust stroke. Gasoline engines us a spark plug to ignite the fuel mixture. Diesel engines have a similar cycle, except that the compression cycle ignites the fuel-air mixture simply because of the increase in temperature that results from the compression. No spark is required. Diesel efficiencies are higher than gasoline efficiencies because the operating temperature is higher and the compression ratio is larger. As a result, diesel efficiencies can approach 60 percent. Truck and bus engines tend to be diesels for this reason. In Europe, many passenger cars are diesels as well because fuel economy is much more important. Diesel power passenger cars are less popular in the United States. Poorly tuned diesel engines emit black soot, which may promote global warming by coating polar ice caps and increasing melting. Soot particles may also be associated with respiratory illnesses, and there is some evidence that they may be able to penetrate the blood-brain barrier.

Less efficient two-stroke gasoline engines are rarely used in cars, but they are still used in chain saws, leaf blowers, lawn mowers, and snow blowers. Two-stroke engines combine intake and exhaust in one cycle. Furthermore, the fuel often includes oil to provide engine lubrication. Two-stroke engines burn fuel less efficiently and emit more pollutants.

Even though the four-stroke engines in cars are more efficient than two-stroke engines, the conversion of fuel energy into kinetic energy in the car is extremely poor. Figure 8.3 illustrates the fraction of the fuel energy that goes into power to the wheels, where the numbers shown are the most optimistic.[3] For city driving, the fraction going to the wheels is only 14 to 16 percent of the energy in the gasoline. Even for highway driving, the percentage is only 20 to 26 percent. A detailed list of the energy loss mechanisms is given in Table 14.

The graphs in Figure 8.3 show that the vast majority of energy is lost to heat, about half of which is due to the fundamental Carnot efficiency limit, governed by the ratio of the engine temperature to the temperature of the outside air. Given that the outside air temperature is not under our control, the loss due to the Carnot efficiency limit cannot be recovered without operating the engine at a higher temperature; however, increasing engine temperatures is technically challenging. Part of the

---

3. United States Energy "Where the Energy Goes: Gasoline Vehicles" provides tabs offering information for combined driving, city driving and highway driving. This information is available at www.fueleconomy.gov/feg/atv.shtml. According to the cite, energy-requirement estimates are based on analysis of over 100 vehicles by Oak Ridge National Laboratory using EPA Test Car List Data Files. The site also offers similar information for hybrid cars.

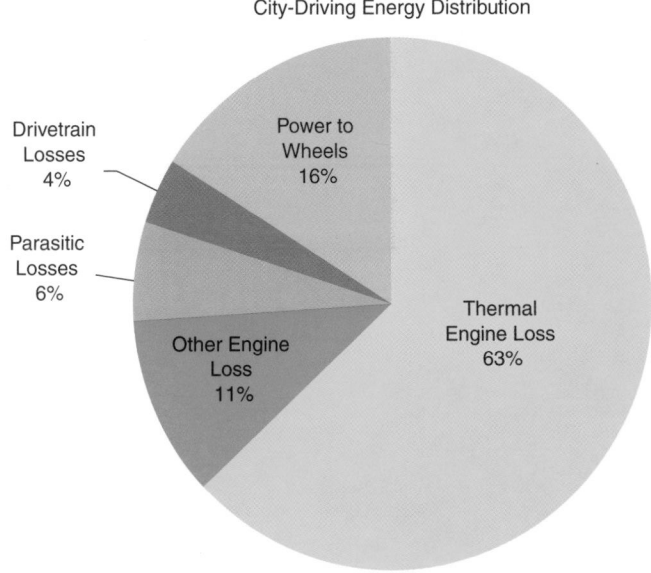

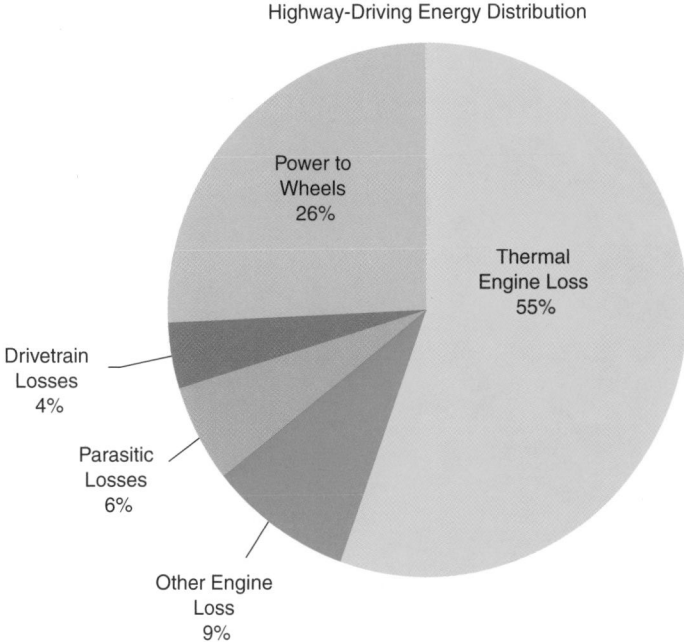

Figure 8.3 Fraction of the fuel energy that is delivered to the wheels, as well as the distribution of the remaining fraction between loss mechanisms for city driving and highway driving. United States Energy "Where the Energy Goes: Gasoline Vehicles" provides tabs offering information for combined driving, city driving, and highway driving. This information is available at www.fueleconomy.gov/feg/atv.shtml. According to the website, energy requirement estimates are based on an analysis of over 100 vehicles by Oak Ridge National Laboratory using EPA Test Car List Data Files. The website also offers similar information for hybrid cars.

Table 14. Detailed energy-loss breakdown for gasoline and hybrid powered cars.

Loss category	Detailed breakdown of contributions	Gasoline engine: city driving (in percent)	Hybrid car: city driving (in percent)	Gasoline engine: highway driving (in percent)	Hybrid car: highway driving (in percent)	Gasoline engine: combined (in percent)	Hybrid car: combined (in percent)
Total engine losses		71–75	66–72	64–69	63–66	68–72	65–69
	Thermal, including radiator and exhaust heat	60–64		56–60		58–62	
	Combustion	3		3		3	
	Pumping	5		3		4	
	Friction	3		3		3	
Parasitic losses	Water pump/alternator	5–7	5–7	3–4	2–4	4–6	4–6
Drivetrain losses		4–5	3–5	4–7	3–5	5–6	3–5
Total power to wheels		14–20	25–40	22–30	29–36	18–25	27–38
	Wind resistance	3–5	6–11	13–19	17–23	9–12	11–16
	Rolling resistance	3–5	6–11	6–9	8–11	5–7	7–11
	Lost braking	7–10	13–20	2–3	3–4	5–7	9–11
	Recovered by regenerative brakes	0	8–14	0	2–4	0	5–9

Source: For comparison, electric cars convert 59–62 percent of their electrical energy into power to the wheels providing 38–41 percent efficiency, http://www.fueleconomy.gov/feg/evtech.shtml.

remaining thermal losses could be improved by insulating the engine better, but such insulation would increase the size, weight, and price of the car. Increases in size reduce gas mileage because of increases in air resistance, and increases in weight reduce gas mileage because kinetic energy is proportional to the mass, so heavier cars require more energy to achieve the same velocity. Automotive design attempts to trade off these different factors to develop a product that will be commercially successful and compliant with all regulations. Fuel efficiency is a consideration in automotive design, but for most cars and trucks it is not the dominant consideration.

Given that so much of the energy loss in internal combustion–powered cars is the result of thermal losses associated with the engines, it is worth nothing that the greatest energy improvements to fossil fuel–burning passenger cars may come not from improving the engines themselves but by making other changes to the cars. For example, decreasing the weight and air resistance of cars can greatly increase energy efficiency. Regenerative braking recycles energy that would otherwise be lost; however, the usefulness of regenerative braking depends on driving style. Similarly, because air resistance increases with the square of the driving speed, the usefulness of reductions in air resistance also depend on driving style. For people who drive predominantly on highways, reductions in air resistance make a big difference. The U.S. Department of Energy offers comparisons of the fuel efficiencies of gasoline-powered vehicles and hybrid vehicles.[4] Those results show that even a person who commutes predominantly on the highway, hybrid cars can offer increases in fuel efficiency since hybrids offer wheel efficiencies of 29 to 36 percent, whereas conventional gasoline cars only offer 22 to 30 percent efficiencies. For a person who commutes 10 miles a day each way at an average speed of 20 miles per hour with frequent starts and stops, regenerative braking may vastly increase energy efficiency, whereas air resistance reductions will have a negligible

---

4. Ibid.

effect. For a person who commutes 60 miles a day each way but spends much of the time stuck in traffic, changing the traffic patterns may produce the largest efficiencies.

In considerations of energy efficiency, it is important to remember that how the devices are used may have a much greater effect on overall energy than small changes in the technology of the devices themselves. For example, more telecommuting or flexible work times could reduce traffic congestion that results in large energy waste because people in cars are sitting still or moving only slowly. Similarly, increasing the number of people walking, biking, or taking public transportation can reduce road congestion. Finally, even if the total number of cars on the road remains the same, self-driving cars that communicate with each other and form efficient convoys might work better than independent decisions made by individual drivers.

Trains

Trains predominantly use diesel engines because of their efficiency, though local trains near major cities often use electric motors to move the trains forward. The principles are the same as those discussed in diesel car engines and electric car motors.

Planes

Propeller planes' internal combustion engines that are similar to those of cars, so they will not be discussed here. Jet planes are not a topic that is much discussed in consideration of renewable energy, possibly because they do not fit into the "all electric" economy as large electrically powered commercial passenger planes are unlikely to ever be developed. However, existing jet planes can run on biofuels; therefore, jet aviation could continue in a 100 percent renewable energy future.

Unlike internal combustion engines, jet engines are not cyclic. Thus, they are not governed by the Carnot efficiency limit; however, they are still quite inefficient. Propeller planes are actually much more fuel efficient than jet plans, but of course they go slower. Technology changes in jet aircraft attempt to modify three major factors: noise, fuel efficiency, and carbon dioxide emissions. All three are driven by economic, regulatory, and consumer forces. Strong economic forces drive engine efficiency development. In 2008, fuel costs represented almost one-third of all airline operating costs. The volatility of fuel prices results in large fluctuations in the fraction of the price that is due to the cost of the fuel. In 2009, fuel costs represented only around 20 percent of operating costs.[5] Enormous energy efficiency increases were observed in the 1960s and 1970s, with smaller efficiencies occurring more recently; however, efficiency increases continue to be substantial. The resulting efficiency gains have been impressive. For example, from 1960 to 2010 the fuel burned per passenger seat has decreased 82 percent.[6] Technology changes play a significant role, and are expected to continue to play a role in the future.[7] Increasing thrust has allowed a transition from four engines to two engines on planes like the Boeing 777. For aviation, the total fuel consumption by the airline has slightly decreased since 2000 despite an increase in the number of passenger miles flown.[8] Incremental engineering improvements in engine design are likely to continue, driven by strong market factors and enabled by advances in materials science and in computing. In addition, real-time computation-based adjustments in engine operation can optimize fuel economy based on what the engine is doing at a particular time. For example, during takeoff and landing, the optimal amount of air bypassing the engine is different than it is when the plane is cruising.

5. The IATA Technology Roadmap Report. Issued June 2009, figure 4-1 p. 34. Available at www.iata.org/whatwedo/environment/Documents/technology-roadmap -2009.pdf. An update is now availale at www.iata.org/whatwedo/environment/Docu ments/technology-roadmap-2013.pdf.

6. Ibid., table 1.2, p. 11.

7. Ibid., figure 1, p. 9.

8. See www.transtats.bts.gov/fuel.asp?pn=0&display=chart1.

Finally, as with many other areas of energy use, the greatest decreases in energy consumption improvements may be obtained by intelligent use management rather than direct technological improvements. Thus, air traffic optimization to avoid circling can greatly improve fuel consumption without any changes in the airliners themselves. Furthermore, scheduling and ticket price optimization can increase the number of passenger miles without increasing the number of flights.

ELECTRICITY AND HYBRID ELECTRICAL TRANSPORT

An electric motor uses electrical current to make a shaft spin around. Such a spinning shaft can move a car forward, raise water from a well, or whip egg whites.

If Sisyphus is the myth governing electric generators, Tantalus is the myth governing electric motors. Tantalus was doomed to stand in a pool of water beneath a fruit tree with low branches without ever being able to eat or drink despite being both hungry and thirsty. Whenever he reaches for the fruit, it moves just out of reach. Similarly, the water always recedes before he can take a drink. The principle behind the electric motor can be explained in terms of carrots and donkeys. In the donkey analogy, a bar is attached to a rotating spindle, and a donkey cart is attached to each end of the bar. As the donkeys walk in pursuit of the elusive carrot, they make the shaft rotate, as illustrated in Figure 8.4. A similar effect can be obtained if a magnet is attached to a rotating shaft, which is exactly what a compass needle is. The compass needle normally moves toward the North Pole of the earth, but if a strong magnet is brought near to the compass, the compass needle will point in the direction of the magnet pole because its field is much stronger than earth's field. The compass needle can be made to spin around by moving the magnet in a circle around the compass, as illustrated in Figure 8.4. If the compass needle were attached to a motor shaft, the turning motor shaft could make a car move forward.

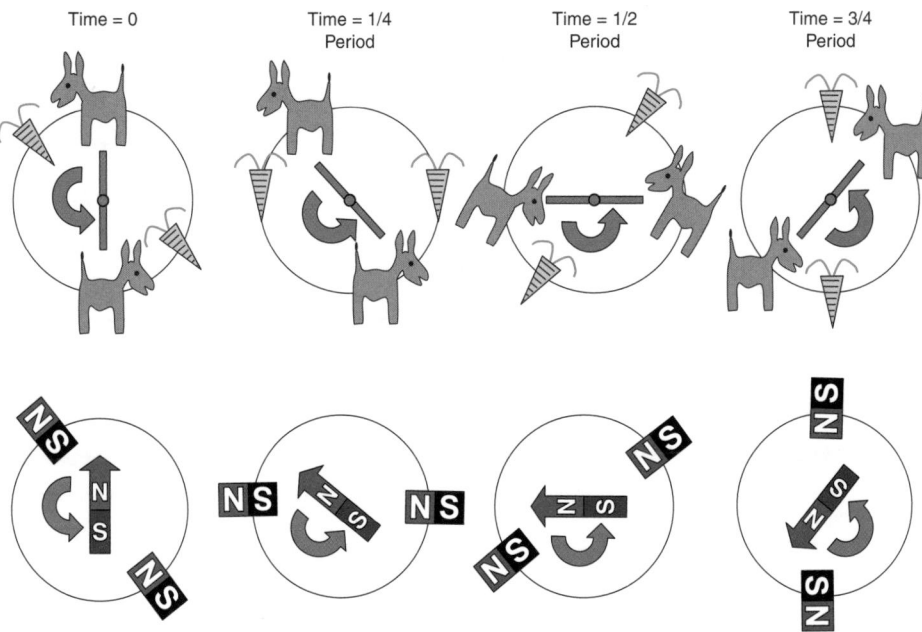

Figure 8.4 Schematic illustration of the fundamental principle behind the operation of a simple electrical motor. The upper diagram indicates two donkey carts connected by a rod. Each donkey is moving forward toward the carrot on a stick which is always just in front of the donkey. The donkeys never catch up to the carrots, so the connecting rod keeps rotating. The lower diagram shows a compass needle that is trying to align with the permanent magnets, where the position of the magnets rotates to keep them just in front of the compass needle just as the carrots remain just in front of the donkeys. The compass needle can never catch up to the magnets, so the rod keeps rotating.

Instead of moving the magnet in a continuous circle, one could be more lazy. Three stationary children, equally spaced around the circle, could play keep away from the compass needle, where each child tosses the magnet to his neighbor just as the needle approaches the child holding the magnet. This is exactly what happens to Tantalus: he reaches for the grapes, and just before he can get to them, they move out of his reach.

Electric motors do not, of course, contain small people passing magnets just before the compass needle reaches its desired position. There are various implementations of electric motors, but one simple type consists of three external electromagnets at the three positions around the circle. If no current flows to an electromagnet, it is not a magnet. Thus, in the absence of current, the electromagnet produces no magnetic field and does not interact with the compass needle. Conversely, if current is flowing, the electromagnet behaves like a permanent magnet. Thus, one can "pass the magnet" by controlling the electric current flowing through the magnet by using switches that are timed so that the current moves to the next magnet just as the compass needle approaches the previous magnet, as illustrated in Figure 8.4. In such a system, one can control the rotation rate of the shaft by controlling the time that the current is switched. This allows infinite tuning of the motor speed. It also allows the motor to run in reverse without changing gears. Thus, electric cars do not require a transmission.

Electric motors provide incredibly efficient conversion of electrical energy to mechanical energy. The efficiency of motors is determined by measuring the electrical power supplied to the motor to the mechanical energy available to turn the shaft.

$$\text{Electrical Efficiency} = \text{(Shaft output power in watts)}/$$
$$\text{(Electrical input power in watts)}$$

There are several sources of loss in electric motors: (1) resistive heating of the wires as the electrons collide with the atoms in the metal, (2) energy lost to heating of the ferromagnetic material used in the electromagnets as the nanomagnets in the material change their direction in response to the magnetic field in the current carrying wires, (3) other losses associated with stray electric and magnetic fields, and (4) mechanical friction in the bearings. Table 15 shows the required efficiencies standards set by the National Electrical Manufacturers Association for class B electric motors.

Large motors clearly have higher efficiency than small motors, but even the lowest efficiency motors perform much better than fossil fuel–burning engines. As shown in Figure 8.3, gasoline-powered internal

Table 15.   Efficiency standards for commercial electric motors.

Motor horsepower at 1,200 rounds per minute (RMP)	Efficiency (in percent)
1	82.5
5	89.5
10	91.7
50	94.1
100	95.0
500	96.2

Note: Based on the National Electrical Manufacturers Association Open 3-Phase Motor Premium Nominal Full-Load Efficiency.

Source: The site also provides information on the annual energy savings that can be expected from replacing older, less-efficient motors with newer, more-efficient motors. For a 200-hp motor, they estimate annual savings in excess of $1000 per year, www.grainger.com/tps/motors_nema _premium_guide.pdf.

combustion engines have conversion efficiencies that are typically below approximately 30 percent, and diesel engine efficiencies are typically around 50 percent. This difference in efficiency explains why converting from internal combustion–powered cars to cars using electric motors would result in such a large improvement in energy efficiency. Such a switch may also greatly reduce emissions, as will be discussed in Chapter 10.

Electric Motors for Transportation

In the last ten years, hybrid vehicles have become much more important. The term *hybrid car* generally refers to a car that includes both an internal combustion engine that burns gasoline to provide energy and an electric motor, which uses electricity stored in a battery to also provide energy. Ferdinand Porsche developed the first electric vehicles in 1898 and followed that up in 1901 with the Lohner-Porsche Elektromobil—a gas-electric hybrid vehicle that was sold in Austria by the Jacob Lohner Company from 1901 to 1906.[9] At present, high-end vehicles, including

9. *History of Hybrid Vehicles* by Brad Berman, June 14, 2011. Available at www.hybrid cars.com/history-of-hybrid-vehicles/.

Porsches, use hybrid systems to provide peak acceleration without requiring large internal combustion engines. What was once the province of deep green environmentalists has now become an accessory of choice for drivers of traditionally high fuel consumption vehicles such as sports cars, pickup trucks, and SUVs.

Whereas internal combustion engines are limited by the Carnot efficiency, electric motors have no fundamental limit and commercially available efficiencies exceeding 90 percent, which is enormous in comparison with the efficiencies of gasoline engine cars as illustrated in Figure 8.3. Mass-market all-electric cars are only beginning to become commercially available, but high-end all-electric cars have been available for some years, including very high performance sports cars. Electric cars offer a number of advantages: (1) much larger energy efficiency, predominantly as a result of having eliminated thermal losses; (2) decreases in emissions; (3) increases in acceleration and (4) decreases in noise, which can be more pleasant for people inside the car but dangerous for pedestrians. The last two effects are independent of the original source of the electricity, but the first two depend strongly on whether the electricity originally came from a renewable source or a fossil fuel–burning source. In addition, though the energy efficiencies of the fossil fuel–burning plants don't depend strongly on the fuel that is burned in the power plant, the emissions do, as will be discussed in Chapter 10.

The subtle advantages of electric motor driven cars include the fact that the motors take up much less space then gasoline engines. In addition, electric cars do not require transmissions, which saves additional space and weight and allows finer real-time dynamic tuning that can increase energy efficiency. Furthermore, electric cars can use hybrid brakes, which recycle the energy that is normally wasted in brake friction, decreasing the amount of initial stored energy required to travel a certain distance.

The battery pack not only can move the car forward, it can provide what is known as "hotel power"—that is, climate control and power for electronics, without the wasteful gas-burning idling required in internal combustion engine–driven cars. All of these factors provide small energy efficiencies, but the large energy efficiency results from eliminating the thermal losses in gasoline-burning engines.

Another way to understand how effective electric cars can be at improving energy efficiency is to describe their performance in terms of "equivalent petroleum mileage." The Department of Energy has defined equivalent petroleum mileage as 82,049 watt-hours per gallon, while driving the electric vehicle over the same urban and highway driving schedules as are used to compute the EPA mileage for other cars, taking into account charging efficiency. The EPA gives electric vehicles an 89-mile per gallon equivalent rating. A crucial piece of this calculation is the efficiency of the conversion of fossil fuel to electricity at the plant and the losses incurred between generation at the plant and the wall plug from which the batteries are charged. This energy efficiency figure is around 30 percent due primarily to Carnot losses. Thus, even a 100 percent efficient electric motor coupled with 30 percent electricity efficiency may be only slightly better than a 20 to 25 percent gasoline engine; however, significant emission reductions can still be obtained if natural gas is used to run the fossil fuel plant. If a nonemitting energy source is used, the emission gains can be enormous.

If electric cars are so fabulous, why do internal combustion engines dominate the world? There are two major reasons: cost and limited mileage range. However, costs are rapidly decreasing and mileage ranges are increasing. The cost of a new Tesla Model S sedan ranges from $70,000 to $90,000, which is significantly larger than most production cars, though comparable to other high-performance luxury cars. Teslas are one of the few cars brands to use lithium ion batteries, which have higher energy densities than any other type of batteries. That high-energy density allows the Tesla to achieve ranges in excess of 300 miles, which is quite practical for most U.S. drivers.

The range of electric cars can be extended by including gasoline or diesel generators that charge the batteries while driving. The Chevy Volt, for example, has an electric motor and a generator but no internal combustion engine. The Volt requires 36 kilowatt-hours per 100 miles and gets 37 miles per gallon in gas-only mode and a 35-mile all-electric range. The Volt gets a 94 miles per gallon equivalent rating, which is approximately the same as the rating for the Tesla.

At present, the "electric car" market is dominated by hybrid systems that have both internal combustion engines and electric motors within one single car. The Toyota Prius is a famous example. Unlike the Volt and the Tesla, which use lithium ion batteries, the Prius uses a sealed 38-module nickel–metal hydride (NiMH) battery pack providing 273.6 volts, a 6.5 ampere-hour capacity, and a weight of 53.3 kilograms (118 pounds). These batteries typically are charged to 40 to 60 percent of maximum capacity to prolong battery life and to provide a reserve for regenerative braking, which will not work if the battery is fully charged.

Though battery performance always degrades over time, the degradation in the Prius is quite graceful. In 2011, *Consumer Reports* magazine tested a 2002 Toyota Prius with over 200,000 miles on it and compared the results with the nearly identical 2001 Prius with 2,000 miles tested by *Consumer Reports* ten years before. The comparison showed little difference in performance when tested for fuel economy and acceleration. Overall fuel economy of the 2001 model was 40.6 miles per U.S. gallon (5.79 liters per 100 kilometers), while the 2002 Prius with high mileage delivered 40.4 miles per U.S. gallon (5.82 liters per 100 kilometers). The cost of replacing the battery varies between $2,200 and $2,600 from a Toyota dealer, but low-use units from salvage yards are available for around $500.

Prius sales are saturating, but this may be due to the much wider availability of other hybrid cars and trucks. Whereas hybrid vehicles were once perceived to be the province of the "Birkenstock-wearing crowd," hybrid vehicles are now manufactured for a variety of market niches. Ford offers a number of hybrid vehicles, as do Lexus and Acura. GM even offers the 2013 Chevy Silverado hybrid pickup truck, which has an EPA mileage rating of 20 miles per gallon in the city, 23 on the highway, and a sticker price of $39,640. Nonhybrid Silverados have prices from $22,000 to $31,000 but have only 15 and 21 miles per gallon city and highway ratings, respectively. The hybrid power train adds at least $8,000 to the cost of the truck. Whether this is a good economic investment depends on how frequently the vehicle stops. The more frequently the vehicle stops, the more effective the investment, as will be explained in the next section. Here, simply note that if the truck is driven 20,000 miles per year in the city and gas costs $4 a gallon, the purchase price difference will be recovered in approximately six years. If the car is driven almost exclusively on the highway, the hybrid feature is of almost no use and the extra weight of the batteries reduces overall efficiency. Thus, for someone who predominantly drives on the highway, the higher purchase price of a hybrid car may never be recovered by gas savings.

Constant solar power charging sounds like an excellent way to power a car. After all, solar-powered photovoltaic planes are able to achieve eternal flight, or at least twenty-four-hour flight through the darkness using batteries to store electrical energy through the night. However, those planes feature large photovoltaic areas and small power requirements. Directly solar-powered cars are unlikely because the energy that can be delivered to an area the size of a car is woefully inadequate; however, they could provide valuable hotel power when cars are stuck in traffic or taxis are waiting for a fare. Thus, the adoption of electric cars depends largely on energy storage technology or on the development of service stations that quickly trade discharged batteries for charged batteries. However, most automobile journeys occur over short distances, where the existing range of electric cars is more than sufficient. It is important to note that for gasoline cars, the energy stored in the gasoline can be sent to the wheels only once, but that for electric cars, the energy can be sent to the wheels once more. At first this seems utterly fanciful, given that conservation of energy is an important basic law of physics; however, the trick is that regenerative brakes allow cars to recycle the kinetic energy of the car, so one given amount of stored energy can be sent to the wheels more than once.[10]

Regenerative Brakes

Regenerative brakes work for exactly the same reason that pulling a swing back and then releasing the swing once allows that child to swing back and forth many times before stopping, even if the child has not yet mastered moving her feet to pump the swing. In the swing system, initial pullback does work against gravity. At the raised height, the child and the swing have gravitational potential energy that is released when the swing is released. As the swing falls, the potential energy is converted to kinetic energy: the swing is lower, but it moves faster. When the swing is at the bottom, there is no stored potential energy, but the kinetic energy allows the swing to rise back up again. In the absence of loss, the child would swing back and forth forever, endlessly converting potential energy to kinetic energy and back again. In the presence of loss, the swing will eventually stop, but the initial potential energy can still allow the child to swing back and forth many times before the losses steal away all of the energy.

Similarly, electrical potential energy stored in a car battery can be used to make a stationary car move. Thus, potential energy in the battery is converted to kinetic energy, analogous to the acceleration of a child as a swing falls, converting gravitation potential energy to kinetic energy. In the absence of friction, once the stored energy was converted to kinetic energy, the car would continue forever without stopping. Even in the absence of loss, eventually most drivers want to stop the car, which is usually achieved by hitting the brakes. Conventional brakes convert the kinetic energy of the car into heat energy in the brakes, as illustrated in Figure 8.5. That heat energy cannot be recovered to move the car forward, so more

10. A nice detailed illustration of how hybrid cars work is available at www.fuel economy.gov/feg/hybridAnimation/hybrid/hybridstarting.html.

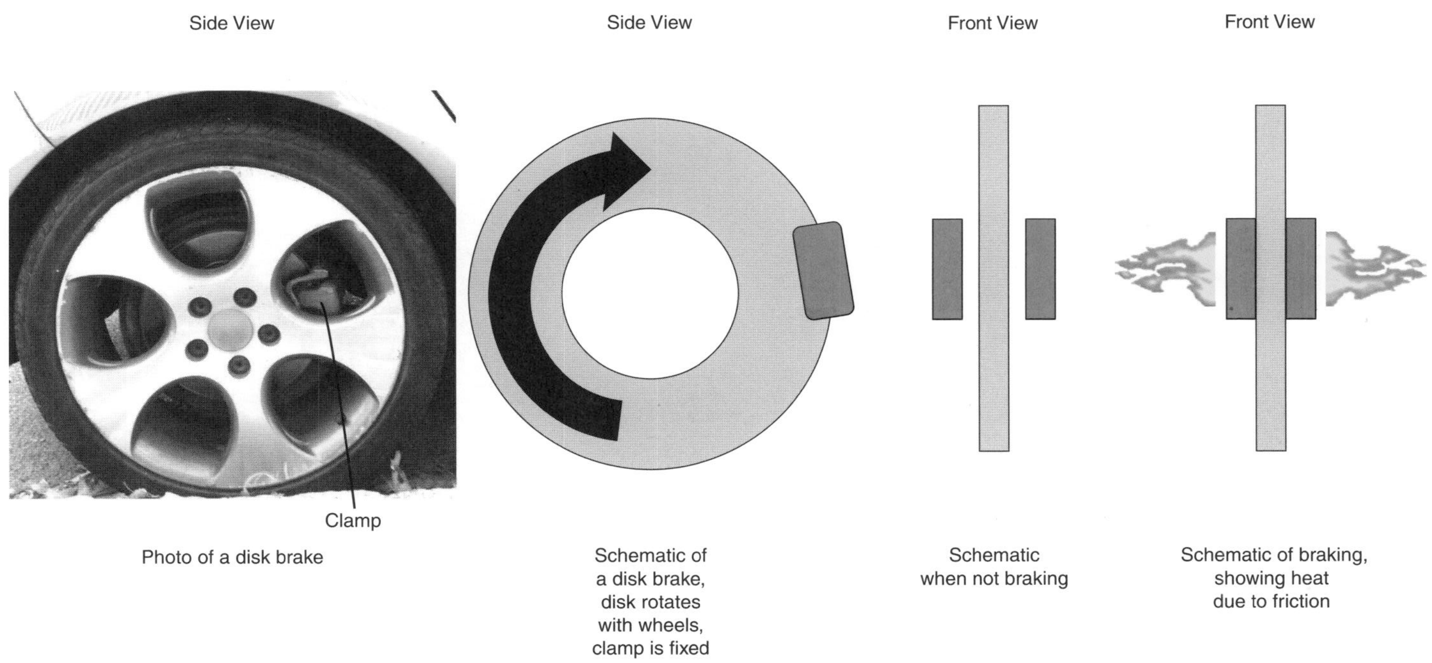

Figure 8.5 Illustration of how conventional disk brakes generate heat when braking because of the friction between the brake pad (red) and the disk (gray). The gray disk rotates along with the wheels. The red clamp is in a fixed position. When the car is not braking, there is no contact between the clamp and the disk. When the brake is applied, the clamp squeezes against the disk, generating friction. The friction stops the car but also heats the disk.

stored energy must be consumed to get the car moving again. Similarly, if at the bottom of the cycle the child puts down her feet, then the friction between the feet and the ground will stop the swing, resulting in warm feet but a swing that will not move until it is pushed again.

The heating in conventional brakes is the result of exactly the same friction that would make the child's feet hot because a conventional brake depends on friction to stop a car. The brake contains a metal disk that is attached to the wheels, and when the wheels turn, the disk turns. When the brake is not being applied, the brake pad does not touch the disk, but when the brake is applied, the pad touches the metal disk and creates friction. Thus, as the pad rubs against the rotating disk, the resulting friction slows the car and the kinetic energy of the car is converted into heat in the brake pad, which is why early automotive brake systems were made of asbestos. In friction-based braking systems, the heat energy that is lost to braking is almost never recovered. Once the car has been stopped by conventional brakes, additional energy must be used to start the car moving again.

In contrast, in a hybrid car with regenerative brakes, the car is not stopped by friction.[11] Instead, when the brakes are applied, the wheels are connected, so that they rotate the shaft in an electrical generator. As a result of Faraday's law, that rotating shaft produces electricity that can be stored in the battery as electrical potential energy. Since the wheels are doing work to create stored electrical potential energy, the car slows down as the kinetic energy is converted into potential energy. As a result, the kinetic energy in the motion of the car is stored in the potential electrical energy in the battery, rather than being wasted in the heat energy generated by frictional brakes. Similarly, swings slow down as their height increases because the kinetic energy present at the bottom of the arc is

11. This is not completely true. Frictional brakes are used once the regenerative brakes have slowed the car down so that it is barely moving forward.

traded for gravitational kinetic energy as the swing climbs back up. Thus, the situation where the regenerative brakes have stopped the car is similar to the situation of the swing at the top of the arc: the swing is not moving because the kinetic energy present at the bottom of the arc has been converted into gravitational potential energy at the top of the arc. Even though the swing is not moving, the kinetic energy has not been lost. Instead, it has been transferred to stored potential energy. That potential energy can be used to move the swing down again, which converts that stored potential energy back into kinetic energy. This process could be repeated forever in the absence of friction. Similarly, the electrical energy stored in the battery by stopping the car using the regenerative brakes can be used to move the car again. In the absence of friction, the car could convert the same energy back and forth between kinetic energy and electrical potential energy forever, just as the swing converts between gravitational potential energy and kinetic energy.[12] Thus, without friction, one single battery charge would power the car forever. This is the magic of regenerative braking.

Overview of Heating, Refrigeration, and Climate Control

Historically, the majority of U.S. residential energy consumption has been devoted to climate control, but that has recently changed. The EIA states that in 2009, 48 percent of energy consumption in U.S. homes was for heating and cooling, down from 58 percent in 1993.[13] The EIA attributes the reduction to increased adoption of more efficient equipment, better insulation, more efficient windows, and population shifts to warmer

12. In reality, even cars with regenerative brakes do a bit of frictional braking.
13. "Heating and cooling no longer majority of U.S. home energy use," U.S. Energy Information Administration," March 7 2013, www.eia.gov/todayinenergy/detail.cfm ?id=10271&src=%E2%80%B9%20Consumption%20%20%20%20%20Residential %20Energy%20Consumption%20Survey%20(RECS)-f1#. Note: Amounts represent the energy consumption in occupied primary housing units.

climates.[14] Importantly, this fractional decrease has occurred along with a decrease in the total energy consumption per household.[15] Though climate no longer represents the majority of residential energy consumption, it still represents approximately half of all energy consumed. Thus, in this section, we will consider energy use for climate control, which encompasses both fully electrical systems and systems that use some combination of combustion and electricity, as well as cases like traditional wood burning fireplaces, which use no electricity at all. I will suggest that combinations of energy sources may offer performance that is superior to all-electric or all–fossil fuel systems. Such combinations offer another opportunity for computation-based real-time adjustments to improve energy performance without lifestyle sacrifices.

Petroleum-burning engines and fossil fuel–burning generators go through cycles, where at the end of the cycle the system returns to its initial state. Engines and generators are subject to the Carnot limit; however, like jet engines, natural gas–burning furnaces, stoves, and fireplaces are not cyclic systems. Thus, natural gas–burning heaters can be more than 90 percent efficient.

Furthermore, though solar heating and geothermal heating may not be widely applicable for electricity generation because it is difficult to obtain sufficiently high temperatures, they may play a significant role in heating and cooling applications where the temperatures required are much lower than the boiling point of water. For example, they may provide heat reservoirs that greatly improve the energy efficiency of heat pumps, which can provide either heating or cooling. Furthermore, rooftop solar water heating systems already play a significant role in warm regions of

14. See www.eia.gov/todayinenergy/detail.cfm?id=10271&src=%E2%80%B9%20 Consumption%20%20%20%20%20Residential%20Energy%20Consumption%20 Survey%20(RECS)-f1.

15. "Residential Energy Consumption Survey data show decreased energy consumption per household," U.S. Energy Information Administration, June 6, 2012, www.eia .gov/todayinenergy/detail.cfm?id=6570.

the world, such as Israel, and they may also play a significant role in the United States.

As shown in Figure 2.1, the largest single U.S. residential electricity use is space cooling, or air conditioning. Of course, the numbers shown in Figure 2.1 represent national averages over a year. Very little air conditioning is used in Anchorage in February, whereas a great deal is used in Miami in July. In areas like the northeastern United States, air conditioning represents a huge load peak in the summer months but a negligible load in winter months. The ratio of the peak load to the average load in New England has become significantly worse over time, increasing from approximately 1.5 in 1993 to 1.8 to 1.9 in 2010–2012.[16] Electric power companies need to maintain sufficient generating capacity to cover those peak demands, but because the full generating capacity is little used, it is highly economically inefficient. If renewable energy sources can level out the peak demand, not only will energy emissions and imports be reduced, but the economics of conventional fossil fuel–burning power plants will also improve because less unused excess capacity will have to be maintained in order to meet consumer demand spikes. Solar photovoltaic power may be particularly helpful in managing peak air conditioning loads since the demand for air conditioning is largest precisely when the solar power generation is largest.

Electric Heaters, Stoves, and Water Heaters

Electric stoves and heaters produce heat for exactly the same reason that incandescent lightbulbs light up: the electrons moving through the heater collide with the metal atoms in the heater. That transfer of energy to the metal atoms heats the metal up. In incandescent lightbulbs, the filaments are often made of tungsten, but electric heaters often use nickel chromium

16. "Peak-to-average electricity demand ratio rising in New England and many other U.S. regions," U.S. Energy Information Administration, February 18, 2014, www.eia.gov/todayinenergy/detail.cfm?id=15051.

wire. Just as electric motors in cars transfer almost 100 percent of the electrical energy into motion, electric heaters transfer almost 100 percent of their energy into heat. From conservation of energy, Q, the heat added is equal to the work done, or in equation form,

$$\text{Heat} = Q = \text{Electrical Work}.$$

Although 100 percent conversion of work to heat sounds ideal, it turns out that heat pumps deliver heat energy that is equal to more than 300 percent of the mechanical work put into the heat pump. This at first seems incredible, but it is true, as will be explained below. Thus, based on energy efficiency alone, heat pumps are preferable to resistive heaters.

Of course, just as in the case of electric motors, the original electricity powering resistive heaters or heat pumps has to come from somewhere. Thus, the expense and efficiency of electric heaters is really a function of the cost and efficiency and the electric power delivered to the heater, so even a so-called 100 percent efficient heater does not actually deliver 100 percent efficiency. Because of the poor Carnot efficiency of fossil fuel–burning electrical generators, electric resistive heating is usually less efficient and more expensive than generating heat locally by burning hydrocarbons. Of course, burning natural gas is favored because it has lower emissions than burning oil.

Despite the low efficiency per unit of heat generated, even electric resistive heaters can result in energy efficiency if they are used in large buildings that do not have zoned heating. Again, it is not just the device itself that governs energy efficiency; it is how the device is used. If heating one room requires burning fossil fuel to heat the whole house, then heating one single room with an electric resistive heater can be more energy efficient. Of course, heating one room using a heat pump powered by renewable energy would be better still. Similarly, using an electric water heater to generate hot water on demand may be more efficient than burning fossil fuel to maintain hot water constantly.

Air Conditioners and Refrigerators

Traditional air conditioners and refrigerators use electrical energy to cool a space. They do it by expending mechanical work, which allows a liquid to constantly evaporate, and that evaporation results in cooling. Thus, cooling occurs for the same reason that evaporating sweat or evaporating rubbing alcohol makes skin feel cold. Molecules in liquids are bound to each other by attractive interactions. This is what makes them stay together as a liquid. In order to convert the liquid into a gas, energy is required to break the bonds between the molecules in the liquid. The required heat energy can come from an object in contact with the liquid, which in the case of sweat or rubbing alcohol is an actual human body. Thus, the evaporating liquid steals the heat energy required to wrest apart the evaporating molecules from the body that is causing the liquid to evaporate. Since the evaporating liquid takes heat away, the body gets colder. Traditional nonelectrical cooling systems simply allow water to evaporate on the roof of a house. No electric power is required, but once the water has evaporated, the cooling ends, unless more water is added.

In contrast, in a refrigerator, the evaporating liquid is contained in pipes, so that the newly formed vapor cannot escape the system. Thus, the evaporated liquid has to be converted back into a liquid and returned to the evaporator, where it can evaporate again. It is this return of the same liquid that makes refrigerators consume electric power, as illustrated in Figure 8.6. The electric power allows the refrigerator to transfer heat from a cold region to a hot region. In the absence of power consumption, this is impossible: heat flows from hot objects to cold ones, making the hot object cooler and the cool object hotter. The spontaneous heat transfer stops when both objects are at the same temperature. The beauty of refrigeration systems is that they transfer heat from cold objects to warm objects, which makes the cold object colder and the warm object warmer. In Figure 8.6, the pink region indicates the warmer environment,

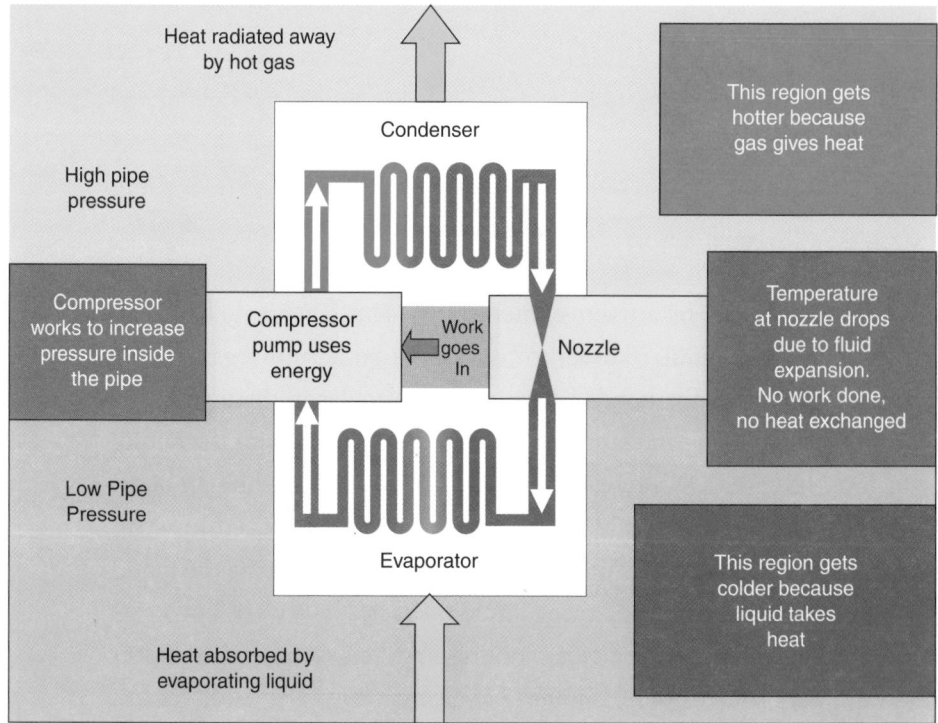

Figure 8.6 Illustration of a simple refrigerator or air conditioner based on evaporation. In the left portion of the graph, the fluid is largely a gas, whereas in the right portion, the fluid is largely liquid. Material leaving the nozzle is in the form of a vapor. In the evaporator, the vapor evaporates, which cools the evaporator coils just as alcohol or sweat cools skin when it evaporates. The evaporated gas then passes through the compressor pump, which increases the pressure of the gas. Work is required to power the pump. That increase in pressure is accompanied by an increase in temperature. The hot gas is now hotter than air outside the condenser. Thus, the hot gas in the condenser radiates heat into the environment, which makes the gas colder. The cool gas leaving the condenser then passes through the expansion nozzle. As the gas passes through the nozzle, its pressure and temperature decrease and the gas is converted into a vapor, which begins the cycle again.

and the light blue region below that indicates the colder environment. The gray region in between indicates the thermal insulation that separates the two so that they cannot directly exchange heat with each other. They exchange heat only through the fluid system. The orange arrows indicate the heat flow, which goes from the cold region into the evaporator and from condenser into the hot region, producing the desired heat flow from the cold object to the warm object.

In the case of a refrigerator in a kitchen, the red region represents the kitchen outside the refrigerator and blue represents the inside of the refrigerator. If the system is functioning as an air conditioner, the red region represents the outside of the house and the light blue region represents the inside of the house, which is cooler than the outside because of the air conditioning.

The system includes the following: (1) a condenser, where the vapor is condensed; (2) an expansion valve, where the pressure of the vapor decreases, which lowers the temperature of the vapor; (3) an evaporator unit, where the vapor evaporates, taking heat away from the cold region, which makes the cold region even colder; and (4) a compressor pump, which increases the pressure and temperature of the gas. The system is quite similar to the steam turbine, except that in the turbine heat was used to generate mechanical work, and in the refrigerator mechanical work is used to move heat from a place that is cold to a place that is hot. Thus, the fluid does work to make the hot region hotter and the cold region colder.

The really strange feature of a refrigerator is that the fluid evaporates or boils in the cold part of the system, while it condenses (i.e., converts from vapor to liquid) in the hot part of the system. The boiling temperature is the temperature at which a liquid converts to a gas. For water at atmospheric pressure, this is 100°C. Similarly, the condensing temperature is the temperature where a vapor condenses into a liquid. For water atmospheric pressure, this is also 100°C, because that is the temperature at which water and steam coexist. Thus, if at atmospheric pressure you have a mixture of steam and water and you neither add nor subtract heat, then you will continue to have a mixture of steam and water forever. If you add heat, you can convert all of the water to steam. If you take heat away, you can convert all of the steam to water. A heat transfer is required to make either change because each change represents a phase transition between the liquid and vapor states of water. During the phase change, the temperature of the system does not change even though heat is being added or subtracted.

This lack of temperature change can be seen if you measure the temperature of water being heated in a tea kettle. Initially, when the stove is turned on, the water temperature rises as the kettle heats. Once the water boils, the temperature remains at 100°C as the water begins to convert to steam.[17] If you keep heating, eventually all of the water will convert to steam, and then the kettle will very rapidly heat up and be damaged. This is why tea kettles have whistle: the whistling occurs because steam is escaping, which means that the water is boiling. While the water is boiling, the temperature does not change because the heat energy is being used to break the bonds between water molecules. The bonds are what connect them to other water molecules. Once the bonds break, they are released as steam. This is the same breaking of bonds that occurs when alcohol or water on skin evaporates, where that breaking requires heat that is taken from the skin, resulting in cooled skin. It turns out that the reverse process occurs as well: if your body temperature were 100°C and steam condensed on you, you would feel warmer. Thus, it makes sense that the evaporator cools because the fluid there is evaporating, and the condenser heats up because the vapor is condensing; however, the condenser is hotter than the evaporator, which is really strange.

If the condenser and the evaporator were at the same pressure, this would be impossible; however, the condenser is at a much higher pressure than the condenser, and the boiling temperature is affected by pressure. The electrical power to the refrigerator is required to establish precisely this pressure difference. To understand how water can boil in the evaporator at a temperature that is lower than the temperature at which the

---

17. It is a general feature of phase changes that the temperature of the system remains constant as long as the phase change is occurring. During that time, the heat being added or subtracted is going into changing the phase rather than the temperature.

same fluid condenses in the condenser, it is useful to consider cooking at high altitudes. In Denver, water actually boils at a temperature slightly below 95°C, though the same water boils at 100°C in Boston. Thus, if one heated water to 99°C in Boston, that water would not be boiling; however, if it were magically transported to Denver without adding or subtracting any heat, the water would evaporate and become steam even if the temperature in Denver were only 97°C. Similarly, steam created at 96°C in Denver would condense and become hot water at 99°C in Boston; therefore, the water condenses in Boston at a temperature above the temperature at which it boiled in Denver. This is precisely the trick in the refrigerator: the evaporator is at a lower pressure, like Denver, whereas the condenser is at a higher pressure, like Boston. Thus, though the condenser temperature is higher than the evaporator temperature, in the condenser the temperature is below the boiling point, whereas in the evaporator the lower temperature is above the boiling point.

Thus, the trick in the refrigerator is transporting the fluid from the low-pressure evaporator to the high-pressure condenser, and back. The transition from the evaporator to the condenser is done using a pump, which consumes electrical energy. The pump does work to compress the gas from the low pressure in the evaporator to the high pressure in the condenser. The compression also warms the gas. One can understand how compression heats gas by considering what happens to a gas molecule bouncing of the piston while it is being compressed. The velocity of the compressing piston provides a little extra push that increases the velocity of the gas molecule above what it would have been if the piston had been stationary. A similar effect can be seen in cheerleading, where the upward push from people in the base allows flyers to ascend faster than they would without the push. As discussed in online Appendixes 2 and 3 at http://thedata.harvard.edu/dvn/dv/HUP, for a gas at a temperature of T degrees kelvin, the square of the average gas velocity is proportional to the temperature. Thus, the push provided by the piston increases the temperature of the gas because it increases the gas velocity.

The high pressure and temperature vapor created by the compressor then passes through the condenser, where it is condensed back into a liquid. After the liquid has condensed in the condenser, it passes through a nozzle, like the nozzle on a water hose. The high-pressure liquid is pushed out into the low-pressure evaporator, just as the high-pressure water is pushed out of the hose nozzle. It turns out that the spreading also lowers the temperature of the fluid. Basically, as discussed previously, water molecules are attracted to each other. Thus, a water molecule represents a potential well for its neighbor. In order to pull two water molecules apart, you have to do work to pull them uphill. In the nozzle, no work is done, but one can envision the water molecules starting with some velocity near the bottom of a potential hill. If the ball is allowed to roll up the hill, it will go more slowly because its kinetic energy is being transferred to potential energy. If it reaches the top of the hill, it will be moving more slowly than it was at the bottom. This is what happens in the nozzle: as the water molecules fly apart, they move up a potential hill, which lowers their velocity, making them colder.

In sum, liquid evaporates as it flows through the evaporator, which cools the evaporator. Work is done to compress the gas to a higher pressure. The high-pressure gas condenses as it flows through the condenser, transferring heat to the condenser. The fluid than flows through the nozzle, resulting in a decrease in pressure and temperature; however, the reduction in temperature is small enough that the liquid evaporates in the evaporator and the cycle begins again. Basically, work has been done to transfer heat from a cold region to a hot region. The work is required because energy spontaneously flows from hot regions into cold ones; reversing the path requires energy.

Heat Pumps Both Heat and Cool

Heat pumps are devices that can either heat or cool. The air conditioner described above is simply the heat pump functioning in air conditioning mode. The same device can act as a heater if the inside of the house is the red region and the outside of the house is the blue region. In that case, the net effect of the mechanical work is to transfer heat from the colder outside region to the warmer inside region. The heat pump is wonderful because it steals heat from the outside air and transports it to the house. As a result, the ratio of heat delivered to work done can exceed one.

One can calculate the ratio of the mechanical work put in by the compressor to the heat delivered to the inside of the house in the limit where no energy is lost and the process is fully reversible. Neither limit can ever be reached, but these assumptions offer a theoretical maximum to which actual results can be compared. In the ideal case, the coefficient of performance (COP), which is the ratio of the heat delivered to the work done, is given by

Ratio of heat delivered to work done $= 1/(1 - T_{cold}/T_{hot})$,   Equation 8.1

where the temperatures are in degrees kelvin. The striking feature of the equation is that it is always greater than one. That means that the heat delivered exceeds the work done, whereas for an electric resistive heater, the heat delivered was at best equal to the work done. As mentioned above, the trick is that not all the required heat is provided by the work done by the compressor. Instead, the heat is "donated" by the colder outside region, which is still "hot" on the kelvin temperature scale, where water freezes at 273 degrees kelvin. The energy paid by the compressor is largely just the energy required to transport the heat from the colder outside to the warmer inside, which is a process that would never happen spontaneously since heat naturally travels from hot objects to cold objects.

A financial analogy is a contest in which the winner receives a "free gift." However, the free gift is not quite free: it requires that the winner pay shipping and handling in order to receive the gift. If the value of the prize is much larger than the cost of the shipping and handling, then the prize winner does indeed win, and it is advantageous to pay the handling fees to receive the much larger value. Of course, if the value of the prize is less than the shipping and handling, then the "winner" has in fact lost. In real life, such financial offers are often scams; however, the scientific version is actually true. The "free" heat comes from the cold outside, which is colder than the inside but still quite warm. When one describes the outside temperature as 0°C, it does not seem that the air outside is warm, but heat energy is not measured in degrees Celsius. The relationship between heat energy and temperature allows individual particles to donate a heat energy that is approximately given by kT, where T is the temperature in degrees kelvin and k is Boltzmann's constant, as discussed in detail in online Appendixes 2 and 3. For one air molecule, the donation is not large, around 10^{-21} joules, but there are approximately 4.4×10^{26} air molecules per cubic meter, so the thermal energy on one cubic meter of air is more than 100,000 joules. On the kelvin scale, water freezes at +273 degrees kelvin, which is a long way from 0 degrees kelvin, even if it does not encourage one to go outside in a swimsuit. Even a −20°F day is balmy on the kelvin scale.

Since the "free" heat is being provided by the cold outside and the work done is the work required to bring the heat in, it is clear that a heat pump will work best when the outside and inside temperatures are nearly the same, in which case the theoretical efficiency approaches infinity. Of course, when the inside and outside temperatures are exactly equal, no heating is required. Though T_{cold} is equal to the outside temperature, T_{hot} is not generally equal to the temperature inside the house. Instead, it is equal to the temperature of the radiators, which is usually higher than the room temperatures since heat must radiate from them into the colder room. Systems that distribute the radiators more broadly throughout the room, such as underfloor systems, can use somewhat lower temperatures than systems with a single radiator at some position in a room. For reasonable choices of temperatures, theoretical ratios are approximately eight;[18]

18. Since the coefficient of performance (COP) is $1/(1 - T_{cold}/T_{hot})$, if the inside of the house has a typical room temperature of 300K and the outside temperature is 273K, the temperature at which water freezes, then the ideal efficiency is 11.

however, what is very impressive is that actual ratios in commercial units can exceed three. That is quite amazing: a real heat pump provides heat energy that exceeds three times the energy that it uses. The COP shows that heat pumps are much more efficient in climates where the temperature differences between the outside and inside are small, whereas in severe climates, the advantage can be substantially smaller (as shown in Equation 8.1). In fact, heat pumps often come with resistive heaters to be used when the outside temperature becomes too low because the heat pump may not be able to deliver enough heat to achieve the desired room temperature since the work it can do is limited.

Thus, though it is clear that heat pump–based heaters are vastly superior to electric resistive heaters, unless the outside temperatures is really low, it is not clear whether they are superior to gas heaters if the electricity required to run the heat pump comes from a fossil fuel–burning power plant. Commercial natural gas heaters can achieve energy efficiencies exceeding 90 percent. Thus, if the electricity from a power plant was only 30 percent efficient, the energy use and carbon emissions from a local gas-fired furnace are approximately the same as those from a heat pump with a 300 percent efficiency. For coal-fired plants, the emission is much worse than the emission from the local gas furnace, as will be discussed in Chapter 10, so using the local gas fireplace would reduce emissions. In addition, from the purely economic viewpoint of the consumer, in many markets the cost of electricity is so much higher than the cost of natural gas that gas heating can be cheaper despite the magical energy efficiency of heat pumps.

For air conditioning, gas burning furnaces obviously do not represent a simple alternative, so for cooling, heat pumps are almost universally used. A calculation of the ratio of the work done to the energy removed from the hot region yields a result that is similar to the case for the heater. The ratio of the work done to the heat transferred is

$$1/(T_{hot}/T_{cold} - 1). \qquad \text{Equation 8.2}$$

Like the COP, which is always greater than one, the efficiency of air conditioners is always greater than one. However, even for nearly identical temperature differences, cooling is less efficient than heating because compressing the fluid provides useful additional heat for the heater that is not helpful for the air conditioner. Again, it is clear that the systems work best when the temperature difference between T_{hot} and T_{cold} is small. Thus, though one has no control of how hot the outdoor temperature becomes, increased efficiency can be obtained by allowing the indoor temperature to be warmer. For very efficient units, this can be a significant contribution; however, for inefficient units, it does not matter as much since most of the energy was lost to something else anyway.

The efficiency of a heat pump can be improved if the "reservoir" temperature is closer to room temperature. Usually the reservoir temperature is the outside air, but on cold days when heat pumps are required to warm houses, the temperature of the outside air will be colder than the temperature of even a shallow geothermal well. Thus, using the geothermal well as the "cold" reservoir may be useful to a heat pump even if the well temperature is lower than room temperature. Similarly, water in a nearby lake or stream can improve air conditioner performance by providing a heat bath that is cooler than the outside air since the lake or stream may be cooler than the outside air temperature at times when air conditioning is required to cool houses.

Heat Exchanges Augmented by Solar or Geothermal Power

In the discussion above, it was assumed that when the heat exchanger is functioning as a heater, the cold reservoir has the temperature of the outside air, where the maximum theoretical energy efficiency is given by $1/(T_{hot}/T_{cold} - 1)$, where T_{cold} is the temperature of the cold reservoir. During winter in the northeastern and northern United States, T_{cold} can be very cold. If a geothermal source is used as the cold reservoir, then the heat pump efficiency can be drastically increased on cold days. Similarly, a cold underground water source, such as a spring, may improve the efficiency of heat exchangers functioning as air conditioners. Of course, direct geothermal heating is also quite feasible and will be considered in a subsequent section.

Solar Water Heating

Residential solar water heating using rooftop systems works well in warm climates and has the added benefit of reducing the heat delivered by the sun to the roof of the house since the solar panels cast shade on the roof and direct heat from the roof to the location where hot water is needed. In such simple thermal solar systems, the sunshine falls on black pipes, which absorb the solar energy. The amount of energy delivered to a given surface area depends on the angle between the surface and the sun. As discussed previously, this changes throughout the day as the earth rotates; however, there is an additional factor that depends on the latitude of the solar panel.

Approximately 4 joules are required to heat 1 cubic meter of water 1°C, so 4,000 joules are required to heat a liter of water by 1°C. U.S. federal regulations limit shower flow to 2.5 gallons or 9.5 liters a minute, so a ten-minute shower requires 100 liters of water. If a square meter of roof is covered by about 5 centimeters of water, then the total volume of water in a square meter is approximately 50 liters, so 2 square meters would be required to hold the 100 liters for a ten-minute shower. Since approximately 4 joules are required to heat 1 cubic meter of water 1°C and 4,000 joules are required to heat a liter of water by 1°C, 8,000,000 joules, or 8 megajoules, are required to heat 100 liters of water to 20°C. If the flux is equal to the maximum solar flux of 1 kilowatt per square meter, then the energy delivered to 2 square meters in one hour is $2 \text{ m}^2 \times 1 \text{ kW/m}^2 \times 3{,}600$ seconds, or approximately 7,000,000 joules. Thus, heating the water for a ten-minute shower would require an hour of sunlight at peak flux. If more roof area were devoted to heating, the required heating time would decrease. Continuous hot water for showering could be provided by around 15 square meters. In most cases, domestic hot water use occurs in surges when people bathe, wash clothes, or wash dishes. Thus, quite small solar collectors with moderate sized storage tanks can provide an excellent supply of hot water for a household. Even at 10 percent efficiency, an average suburban house roof easily accommodates the hot water needs of a household. In 1980, the Israeli Knesset passed a law requiring the installation of solar water heaters in all new homes (except high towers with insufficient roof area). As a result, Israel is now the world leader in the use of solar energy per capita, with 90 percent of the households today using solar thermal systems (4 percent of the primary national energy consumption), estimated to save the country two million barrels of oil a year.[19]

Geothermal Heating

In Iceland, 90 percent of heating energy is provided by geothermal sources.[20] Geothermal heating and cooling systems are used in the United States as well. In the United States, average installation costs range from $22,000 to $30,000 for a typical new home.[21] Given that average heating and cooling costs in the United States are $1,000 to $3,000 per year, the geothermal investment will pay off in ten to thirty years.[22]

Larger-scale geothermal climate control systems are also being built in the United States. On August 18, 2010, Maine's *Portland Press Herald*

19. Solar power in Israel available at en.wikipedia.org/wiki/Solar_power_in_Israel, retrieved July 4, 2014.

20. Icelandic National Energy Authority Direct use of geothermal energy, available at www.nea.is/geothermal/direct-utilization/nr/91. Iceland also generates approximately 25 percent of its electricity from geothermal sources, www.nea.is/geothermal/.

21. See www.energyhomes.org/renewable%20technology/geoinstallation.html. Opportunities for individual calculation are available at www.geothermalgenius.org /thinking-of-buying/average-cost-of-geothermal-heat-pump-installation.html.

22. "Heating costs for most households are forecast to rise from last winter's level," U.S. Energy Information Administration, October 8, 2013, www.eia.gov/todayinenergy /detail.cfm?id=13311.

discussed geothermal heating and cooling that will be included in the new Portland jetport. The system will warm and cool the jetport's 137,000-square-foot addition. The system will cost $3 million. Roughly $2.5 million is coming from a federal program aimed at reducing air pollution and climate change emissions at airports. The balance will come from fees paid by airport passengers. Though Maine is not a major user of geothermal power, commercial systems are being used by schools, including the Gorham Middle School, the University of Maine at Farmington, the University of Southern Maine in Portland, and Bowdoin College in Brunswick.

Burning Fossil Fuels

Household furnaces heat homes by burning natural gas or oil to create heat. Issues with Carnot efficiency do not exist because these furnaces do not cycle a fluid between reservoirs at two different temperatures. They simply release chemical energy by burning. The energy released is the product. Existing commercial boilers and gas-fired home furnaces already have efficiencies exceeding 98 percent, so substantial efficiency improvements are nearly impossible. Of course, existing commercial heat pumps can provide more than three times as much heat as they consume electrical energy if the temperature difference between the inside and outside is small. Whether heat pumps are advantageous depends entirely on the primary energy source used to produce the electricity that powers the heat pump.

LOSSES FROM CONVERSIONS INTO HEAT

Most energy losses that have been considered so far in this book are dominated by fundamental limits, such as the Carnot limit and the Betz limit, where many existing systems already approach the fundamental limits. Frictional losses are not governed by a similar simple fundamental limit, but losses associated with friction cannot fully be eliminated.

In this section, I will consider reductions in friction that may produce significant energy efficiency improvements in lighting and transport. Dramatic decreases can be obtained by switching from processes that require friction to processes that do not. The great energy improvement that results from the shift from frictional braking to regenerative braking has already been considered. In this section, I will consider the similar shift from friction-based incandescent lighting to non-friction-based LED lighting.

In the Transmission of Electricity along Wires

The majority of the energy losses go into waste heat. Some "waste" heat is recycled and used in other processes requiring heat, but most is simply lost to people forever. Similarly, joule heating in wires due to the collisions between electrons and metal ions reduces the efficiency of long-distance electricity transmission. As discussed in Chapter 7, these losses in long-distance transport are being reduced by increasing the voltage at which electrical power is transmitted. Furthermore, if science finds a way to create room-temperature superconductors that can transmit high currents, the frictional losses could be eliminated altogether; however, such losses represent only about 7 percent of the total U.S. electricity use, so even eliminating all losses would not make a tremendous difference to energy consumption.

In Transportation

In transportation, there are several types of frictional losses. Figure 8.7 shows the percentage of the frictional losses for passenger cars that result from braking, rolling friction, and wind resistance. For hybrid cars or electric cars, the losses due to braking can be greatly reduced, but they are of little use in cars with only internal combustion engines. Unsurprisingly, in city driving the losses are dominated by braking, which is why hybrid cars can be very energy efficient when driven in cities. Also unsurpris-

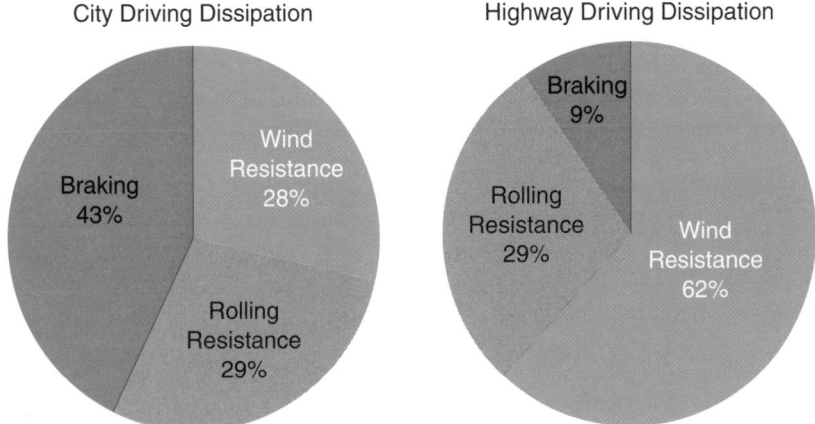

Figure 8.7 Energy lost to different dissipation mechanisms in city driving and in highway driving. Rolling resistance is approximately the same in both, but in city driving, the losses are dominated by braking, whereas in highway driving, the losses are dominated by wind resistance, which includes both skin friction and form drag. The difference in losses owing to braking makes hybrid cars with regenerative brakes much more efficient in city driving but not much more efficient in highway driving.

ingly, in highway driving the loss contribution to braking is small, which is why hybrids are not helpful in highway driving.

Rolling resistance results from energy losses associated with the compression of the tires as the car moves. If you look at a car from the side, you can see that the tires are slightly flattened where they touch the ground. That flattening represents a distortion in the shape of the otherwise round tire. Work is required to distort the shape, as you can readily determine by kicking a tire. If the tire bounced back perfectly, the work done to compress the tire would be recovered as the tire rotated; however, rubber does not perfectly bounce back, as anyone who has dropped a rubber ball knows. The energy that does not bounce back is lost to heat. The problem is reduced if the tire is more inflated, so that the rubber distorts less. Noncompressible tires would not suffer from this loss, but driving around on the rims has limited appeal. Unlike braking and air resistance, which depend strongly on whether a car is being driven in the city or on the highway, Figure 8.7 shows that the fractional loss due to rolling resistance is similar for city and highway driving.

The last loss mechanism is wind resistance, which is the dominant loss mechanism in highway driving and is very high because the power loss due to collisions between the vehicle and air molecules is proportional to the cube of the velocity. This loss mechanism has exactly the same origin as the cube of velocity scaling in wind power. In the case of wind power, the kinetic energy in the moving air is transferred to the stationary wind turbine because the moving air molecules hit the turbine. To a person riding at the speed of the wind, the air molecules would appear stationary, but the windmill would appear to be coming toward the person at a velocity equal to the wind speed. This effect is familiar to passengers in moving cars or trains: as they look out the window, they see stationary objects attached to the earth moving toward them at the speed of the car.

To a person in the moving car, the stationary air molecules would seem to be approaching the car at the speed of the car. Thus, from the perspective of the car passenger, the car is experiencing a wind with a speed equal to the speed of the car. This wind can be felt by anyone who sticks his hand outside of a moving car, and it can be seen when a dog sticks his head out of the car and has his fur blown dramatically back. Each collision between the car, hand or dog hair, and an air molecule transfers energy from the car to the air, which is pushing back against the motion of the car, where one can clearly feel that larger speeds are associated with much larger pushes.

In either the wind turbine or the moving car, each collision with an air molecule results in an energy transfer that is proportional to the square of the velocity, which results in a power transfer that is proportional to the cube of the velocity. In the case of motor travel, higher velocities also result in faster trips, so the total energy lost during a trip over a distance d is proportional to the square of the velocity. Thus, a vehicle moving

55 miles per hour suffers half the air resistance of a vehicle moving at 77 miles per hour. This is why the national 55 miles per hour speed limit was instituted during the energy crisis in the mid-1970s. Although the reduced speed limit made technical sense, it was enormously politically unpopular. It is not clear whether it would have been more popular if people had had a clearer understanding of air friction.

Air friction is usually divided into two categories: skin drag and form drag. Form drag is due to the transfer of kinetic energy to the air molecules because the molecule hits the object. Form drag would be present even if the air molecules did not interact with each other. In contrast, skin friction is due to the viscous drag that results from air molecules that do not directly contact the car. Usually, we don't think of air molecules as interacting, but they do interact because they are attracted to each other. Since molecules are attracted to each other, pulling on one air molecule also pulls on its neighbors. You can feel this effect by positioning your nondominant hand perpendicular to the floor with the palm facing up. If you then place your dominant hand with its palm perpendicular to the floor and just above your nondominant hand, by waving your dominant hand back and forth, you create a breeze that you can feel on the palm of your other hand. It is easier to feel the effect if your stationary hand is wet.

In contrast, if the air molecules did not interact, the motion of one molecule would not affect any other. In that case, you would feel a breeze only if you moved your hands toward each other in a clapping motion where your one hand sent molecules directly to your other hand. Thus, if a car starts moving, it will pull on air molecules that are positioned beside the car just as your hand pulled along air molecules that were not directly in front of your hand. These molecules just next to the car are literally torn since molecules far above them are still, but molecules in the path of the car move at the speed of the car. Molecules just above the car can either match the speed of the molecules near the car, or they can match the speed of molecules far away where the air is stationary. They cannot do both. In fact, they compromise and choose a speed partway between the speed of the vehicle and the speed of distant molecules. As a result, a moving car not only gives energy to molecules that directly hit the car; it gives energy to molecules that are near to the car. The contribution to the drag from molecules near the car is called skin friction.

Racing cars have very low profiles, so they hit very few air molecules directly; therefore, for them air drag is dominated by skin friction. In contrast, vans and SUVs have much higher profiles, so they hit a lot of air molecules directly. Thus, for them the air resistance is dominated by form drag.

In general, the total force in a vehicle is given by

$$F = 1/2 \times \rho \times v^2 \times C_d \times (\text{effective area}), \qquad \text{Equation 8.3}$$

where ρ is the mass density (mass/volume) of air and C_d is a drag coefficient. "Effective Area" is a nebulous concept. A derivation based on the wind turbine example would have arrived at a factor that is equal to the actual area; however, drag is a bit more subtle because shape has a large impact on form drag coefficients. Shapes that allow the air to pass smoothly around the object have smaller air friction than square shapes that deflect the air up because momentum conservation implies that the loss of forward air motion results in a backward push on the moving object. This problem with square shapes is what has led truckers to add curved shields that connect the square top of the cab to the top of the truck. The shield deflects the air smoothly around the trick rather then deflecting it straight up, which reduces the air drag and improves gas mileage.

In practice, the drag coefficient for cars and planes is measured experimentally in wind tunnels. At present, cars have typical drag coefficients of around 0.3 to 0.4. Hummers, with their boxy square profiles, make a statement with their drag coefficients of around 0.6. Ford Mustangs and Toyota trucks have coefficients of around 0.4, whereas Teslas have coefficients of about 0.25. Concept cars have coefficients of less than 0.2, and highly optimized vehicles have coefficients of less than 0.1. Thus, just changing the shapes of vehicles could improve gas mileage or electric car driving range by a significant factor, without requiring a significant lifestyle change as long as one is not emotionally attached to boxy car shapes.

In sum, braking losses cannot be reduced for internal combustion cars, but regenerative braking can greatly reduce the loss for hybrid electric or electric cars. Rolling friction can be reduced by increasing tire inflation, as pointed out by President Obama—unfortunately, that announcement seemed to do more to increase hot air resulting from political commentary than to have stored air resulting from increased tire pressure. Finally, air friction can be reduced by driving more slowly. This is a dramatic effect as the loss is proportional to the square of the speed, meaning that making a trip at 80 miles per hour requires four times as much energy as making the same trip at 40 miles per hour, but few people have the patience to drive more slowly. However, streamlining can reduce losses without requiring slower driving, which is why cars are becoming increasingly streamlined.

Radiation

AC electrical systems radiate energy resulting in losses in transmission and power conversion. Power companies are acutely aware of this issue, so incremental improvements will probably continue; however, the losses are not so large (less than 10 percent) that completely eliminating them would be transformative.

If two objects at different temperatures coexist in the same environment, the hotter object will transfer energy to the colder object, as can be demonstrated when one's hands become chilled or warmed when holding either a glass of cold soda or a mug of hot tea. The rate of transfer depends on how well thermally isolated the objects are. Improving insulation in old uninsulated buildings can vastly improve their energy efficiency. New buildings in the United States tend to be much better insulated than older buildings, but Europe has a much higher emphasis on building insulation in new construction than the United States does.

Smart buildings not only use better insulation, but they distribute energy within buildings so that it is used only when required. If the family is spending the day on the first floor, considerable energy savings can be obtained by not heating or cooling the second floor. This is loss reduction due to not consuming energy that is not truly useful to the consumer. Combinations of insulation, efficient energy conversion systems, and smart management allow state-of-the-art buildings to use approximately 50 percent of the energy of conventional buildings.[23] Importantly, these reductions can be achieved at almost zero cost since the cost savings due to energy use reductions are approximately equal to the additional costs of the energy saving features.[24] As discussed previously, such incremental changes in buildings have allowed the residential energy consumption associated with climate control to drop below half of the total energy used, which contributes to the overall decrease in energy use per household that has occurred in the last ten years.

23. "Real Prospects for Energy Efficiency in the United States," The National Academies Press, sec. 2.4.1, p. 62. 2010. Available at www.nap.edu/catalog.php?record_id=12621: "a small but growing subset of new commercial buildings achieve a savings of 50 percent." (relative to prevailing model Energy Code ASHRAE 90.1). On page 63, the work goes on to state that "the net incremental first cost of achieving a 50 percent reduction in energy" use through an integrated approach can be at or near zero; the savings from downsizing and simplifying HVAC systems generally pay fully for the additional costs of measures such as additional insulation, better windows, and daylighting." In general, this document has a wealth of information about possible increases in energy efficiency. This document includes detailed consideration of the costs associated with these choices.

24. Ibid., 63.

9

STORING ENERGY

Energy storage has become enormously important during the last fifty years and is likely to become even more important as wind and solar power make more significant contributions to electrical power. The explosion in consumer electronics has been a major driver for energy storage systems. Laptops, cell phones, and similar electronics depend on batteries to provide portable energy sources. In those systems, size and weight are extremely important. No one likes a fat heavy cell phone.

For a given amount of energy, the weight of a storage device is determined by the energy per unit mass, which in SI units is measured in joules per kilogram. Similarly, for a given amount of stored energy, the size of the storage device is determined by energy per unit volume, which in SI units is measured in joules per liter. Table 16 provides typical energy densities of various storage systems, where both energy per mass and energy per volume are provided. These numbers must include not only the fuel itself but any packaging required to confine the fuel. A lead acid battery needs to include the plastic case, for example. Gasoline and diesel are more than ten times better than lithium ion batteries for both values. This is the fundamental problem with conversion to electric cars since energy storage severely limits their range.

Table 16.   The energy storage properties of different systems.

Storage system	Energy per mass (kg)	Energy per volume (liter)
Gravitational (potential energy behind 100-meter dam)		
Water Stored Behind a Dam	0.001 MJ/kg	0.001 MJ/L
Chemical (atomic electrical potential energy)		
Nonrechargeable battery	2 MJ/kg	4 MJ/L
Rechargeable battery	1 MJ/kg	1 MJ/L
TNT	4 MJ/kg	
Ethanol	26 MJ/kg	24 MJ/L
Gasoline	46 MJ/kg	34 MJ/L
Hydrogen gas	140 MJ/kg	0.01 MJ/L
Nuclear		
Fission (U-235)	9×10^7 MJ/kg	2×10^6 MJ/L
Fusion (H-D)	3×10^8 MJ/kg	
Mass energy		
Matter antimatter annihilation	9×10^{10} MJ/kg	

Note: Antimatter annihilating in a reaction with matter is best in all categories; however, it has no practical value. Hydrogen gas has a lot of energy per unit mass, but it has a low energy per liter, and handling it can be tricky. Thus, in practical terms, gasoline and diesel are the best energy storage systems, consistent with the fact that they continue to dominate transport.

How long a battery-powered device will run depends on how much energy is stored in the battery. The battery-driven mechanical toy bunny will run longer if the amount of stored energy is larger. Similarly, the stored energy determines how far an electric car can go before a recharge, or when one's cell phone will give out, or when the laptop will no longer function on a plane. Thus, it is a very important parameter. However, there is also a second very important parameter: the power that can be delivered. Power is energy over time. Thus, power determines whether the bunny can put the drum stick through the drum or how high the bunny can jump. Capacitors supply lots of power but don't store much energy. In contrast, batteries store a lot of energy but cannot provide much power.

POTENTIAL ENERGY–BASED STORAGE

The vast majority of both natural and artificial energy-storage systems contain the energy in a potential difference between two different points in space. In reusable systems, work is done to move particles up a potential hill. In gravitational systems, work is done to lift a weight against gravity, creating gravitational potential energy. The stored energy is released when the weight is allowed to fall back down. Similarly, in most electrical systems, work is done to separate oppositely charge particles that are attracted toward each other. The stored energy is released when the particles are allowed to move closer together. In contrast, non-reusable systems like fossil fuels and nuclear fission systems involve creating conditions in which charged particles that were trapped in a high-potential energy state can move toward a lower energy state, releasing energy in the process.

Gravity

Large-scale electrical power storage is dominated by systems that use dams. These systems can have efficiencies of around 90 percent. During periods where energy is plentiful, an electric pump is used to pump water from a lower reservoir to an upper reservoir. During times where stored energy is required, the stored water flows past the generator, just as it does in conventional hydroelectric systems. The storage time for such systems is unlimited, though there are losses associated with evaporation of the water. Similarly, the system can be used an infinite number of times without fundamental degradation, though mechanical parts and physical plant have to be maintained. Pumped hydropower is many orders of magnitude superior to any other system if one considers discharge time at rated power. In a hydroelectric system, the available power is limited by either the generators or the allowed water flow. Most locations do not offer two large reservoirs at different heights; however, in places where a large natural reservoir is available, a nearby artificial reservoir can be built to provide water storage even if a natural higher reservoir is not available.

One example of an artificial pumped hydro facility is the Ludington Pumped Storage Plant. The plant was built between 1969 and 1973 at a cost of $327 million.[1] The artificial reservoir is located on the banks of Lake Michigan, which supplies water for the reservoir. The reservoir is 2.5 miles (4 kilometers) long, and one mile (1.6 kilometers) wide and holds 27 billion U.S. gallons (100,000,000 m³) of water.[2] The power plant consists of six reversible turbines that can each generate 312 megawatts of electricity for a total output of 1,872 megawatts, which is enough to serve ~1.4 million people. An upgrade is expected to increase the capacity by approximately 15 percent by 2020. Water is delivered from the upper reservoir to the turbines by six penstocks, each of which is 1300 feet long and 28 feet across.[3] When electricity demand is low, the turbines pump water 363 feet (110 meters) uphill from Lake Michigan into the reservoir.[4] During periods of peak demand, water is released to generate power. The Ludington Pumped Storage plant is connected to three 345-kilovolt transmission lines.[5]

Efforts are now being made to commercialize systems that store energy in the gravitational potential of weighted railway cars that are towed up tracks like roller coasters.[6] The stored energy in such systems is given by the product of the mass of the object times the gravitational constant

---

1. See www.consumersenergy.com/uploadedFiles/CEWEB/SHARED/Ludington PumpedStorage, PDF page 8.

2. See www.visitludington.com/stories/ludington_pumped_storage_project.

3. See www.consumersenergy.com/content.aspx?id=6985.

4. Ibid.

5. The existence of the power lines provided an incentive for a wind farm to be located nearby, since the wind farm could exploit the existing power lines. See www.grbj.com/articles/77441-consumers-dte-pump-up-energy-storage.

6. "Energy storage: Packing some power," The *Economist Technical Quarterly* Q1 2012. From the print edition, March 1, 2012. Available at www.economist.com/node /21548495.

times the height. To store one megajoule of energy requires approximately a ton of material stored at a height of 100 meters. A megawatt-generating power station uses a megajoule of energy per second. So, one minute would require 60 megajoules of energy, and an hour would require around 3,000. That is some roller coaster.

Electrical Storage

At present, portable energy storage is dominated by electrical potential–based storage. This includes the following: (1) fossil fuels that use burning to release energy stored in the electrons surrounding atoms by allowing the electrons, which are negatively charged, to rearrange themselves so they end up closer to the positively charged nuclei to which the electrons are attracted, (2) batteries that store and release the energy stored in electrons surrounding positively charge nuclei by changing the separation between electrons and the positively charged nuclei, where the distance increases when energy is stored and decreases when energy is released, (3) fissionable materials, which release the electrical potential energy stored in nuclei by allowing positively charged nuclei to rearrange themselves so that they end up farther apart after the reaction, and (4) capacitors that store charges on separate conducting plates without significantly altering the electronic structure of the conductors.

Nonreusable Storage

Fossil fuels and fissionable materials contain electrical energy that can be released, but we do not take another form of energy and convert it to fossil fuel– or nuclear-based storage. Similarly, "primary batteries" provide stored energy for one use, but they cannot be recharged. Thus, these systems cannot be used to store energy generated by windmills or solar systems. They are also useless in electric or electric hybrid cars, and primary batteries, though often used in cameras and toys, are rarely used in portable computers or cell phones.

In burning reactions, the oxygen from the air combines with the material being burned to create an oxidation reaction. Thus, fossil fuels have an advantage in energy storage density because oxygen required for their reaction is not stored in the fuel itself. Furthermore, natural gas and petroleum burn almost completely. A gas tank has nothing in it when the fuel is consumed. In contrast, when a battery is empty, a large object is left over. That large object must be disposed of in some way, which represents an energy use issue that is not frequently discussed. The $50 billion global battery market is dominated by these "primary" batteries that must be thrown away after they are used once. Rechargeable or "secondary" batteries account for only around 10 percent of battery sales.[7] Of course, "rechargeable" batteries eventually need to be disposed of as well, but they have typical lifetimes ranging from three to ten years and can accommodate thousands of charging cycles. Thus, using rechargeable batteries tremendously reduces the volume of waste that must be managed.

Reusable energy storage systems, such as rechargeable batteries and capacitors, allow energy in one form to be stored as electrical chemical energy and then released at a later time when the need for the energy arises. Thus, when people discuss energy storage, they usually mean reusable storage; however, for examples like motor vehicles, the excellent energy density provided by gasoline and diesel allow them to dominate the energy storage market. Similarly, primary batteries have higher energy densities, so they are preferred to rechargeable batteries in some applications where long lifetime is of great importance and the battery use is not heavy.

Reusable Storage

Non-reusable storage, including fossil fuels, has dominated energy storage; however, the portable electronics revolution created tremendous demand for systems that can store a lot of energy in a small light package.

---

7. See batteryuniversity.com/learn/article/battery_statistics.

As a result, enormous time and effort have been and continue to be invested in renewable storage. The advent of hybrid and electric cars has created an additional market for reusable storage, which in some cases exploits advances in the batteries that power portable electronics. One reason Tesla motors was able to bring an electric car to market so quickly was that the batteries that they use are simply groups of batteries designed for laptop computers. Finally, the intermittent nature of wind and solar power creates additional incentives for research on reusable storage systems. One approach is to use distributed storage. For example, each house with a solar panel could also have a storage system that accumulates energy while the sun is shining and returns it to the house when the sun is not shining. It has been suggested that the battery packs used for electric cars could provide such storage, and a quick calculation shows that the stored energy in existing battery packs is more than sufficient for such an application.[8] This possibility may be one the motivations for the creation of a new battery gigafactory.[9] This application is one of the drivers behind the other approach—to have very large scale, "grid-scale," storage that could store energy for large numbers of consumers.

Capacitors

Capacitors in their simplest form consist of two metal plates. Initially, both of the plates have equal numbers of positive and negative charges, so they are electrically neutral, as illustrated by the first drawing on the left in Figure 9.1. Given that they are electrically neutral, there is no average force acting on the electrons, and there will be no net motion of

8. Tesla's come with 60 kWh or 85 kWh batteries. An average household uses approximately 1 kilowatt of average power, so 24 hours of storage would be 24 kilowatt hours. Thus, existing Tesla batteries have much more storage capacity than would be required for an average household to use solar power to disconnect from the grid. See www.teslamotors.com/en_EU/models/features#/battery.

9. "Elon Musk's gigafactory: Assault on batteries," The *Economist*, June 12, 2014, from the print edition. Available at www.economist.com/news/business/21604174 -better-power-packs-will-open-road-electric-vehicles-assault-batteries.

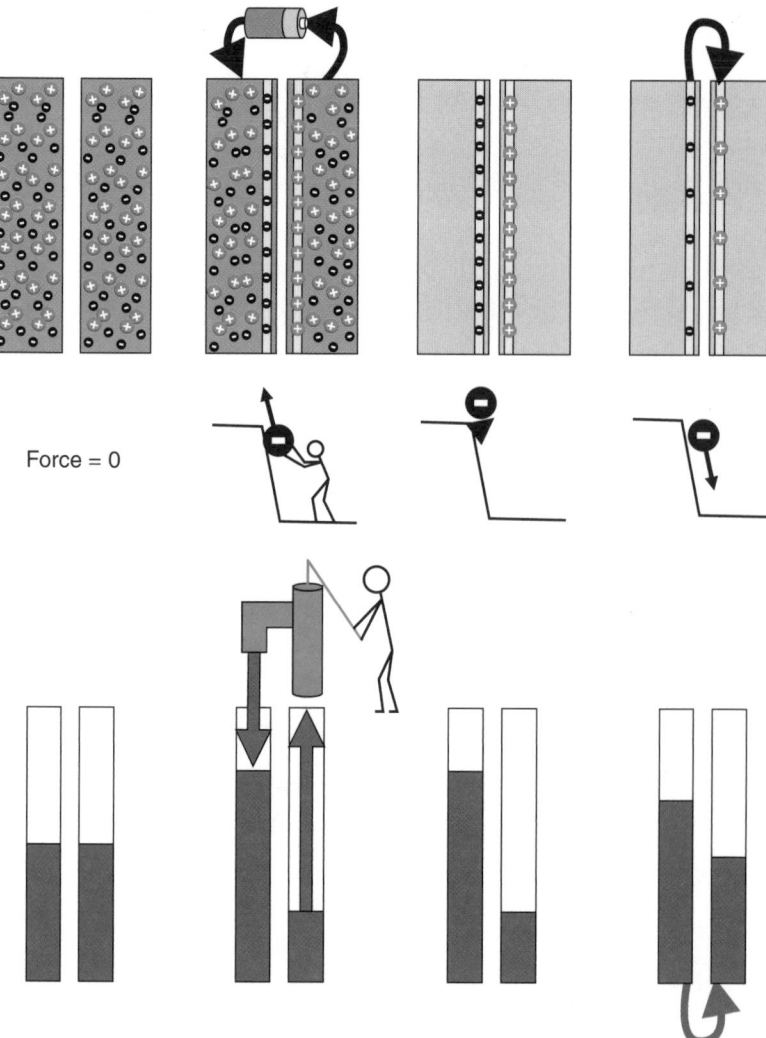

Figure 9.1 Illustration of the basic principle behind energy storage in capacitors. The top panel represents a real capacitor. The middle panel represents a gravitational analogy based on pushing a boulder uphill. The bottom panel represent a gravitational analogy based on two neighboring water wells. Vertical columns represent that same stage in charging. In the first column, the two capacitor plates have no net charge, though both plates contain large numbers of free charge, as shown by the red circles representing the positively charged nuclei and the black circles representing the free electrons. Even though there is lots of charge, the among of charge on both sides is equal; thus, there

(Figure 9.1 continued) is no advantage for a charge to move from one side to another. Moving a single charge from one plate to another requires energy. When a battery is applied, the battery begins to pump electrons from one plate to the other. As a result, the plate to which the electrons are pumped acquires a negative charge and the plate from which the electrons are pumped acquires a positive charge. Moving more charge requires additional energy because the excess electrons push back against any new electrons being added. Thus, the capacitor plate containing the negative charges is located up an electrical hill from the plate containing an excess of positive charge. In the capacitor, the excess negative and positive charges are located at the inner surface of the capacitor because they are attracted to each other. The rest of the capacitor is electrically neutral, as indicated in the two left-hand columns, where electrically neutral regions are shown in gray to highlight the positions of the excess charges in each plate. In the water analogy, the pump is doing work to move the water uphill and place it at a higher level. If the battery is removed, the charge will remain on the capacitor, at least for a while. If the two capacitor plates are connected, the excess negative charge will flow back toward the positively charge plates until both plates have no net charge again. Similarly, if the two water wells are connected, the water will flow from the higher well to the lower well until both wells again have the same level.

---

charge. If battery or other voltage source is connected between the plates, the battery will push electrons from one plate to the other, leaving one positively charged plate and one negatively charged plate. This creates an electric potential hill between the two plates. If the power source is disconnected and no wire connects the plates, the potential hill exists, but no charge will flow. When a wire connects the plates, the electrons travel through the wire to the positively charged plate. Those flowing electrons can do work. This system works remarkably well. It offers tremendous power. It can be charged and discharged indefinitely without any loss of storage capacity. There is no fundamental limit on how quickly a capacitor can charge or discharge; thus, capacitors can store or delivery tremendous amounts of charge in a very short time. Energy over time is power, so capacitors are used in high-power applications. For example, they were used to provide the high power that produced exploding flash lighting in cameras.

One problem with capacitors is that if one tries to store too much charge, the charges simply leap from one capacitor plate to the other. In effect, it creates lightning inside the capacitor. A second problem is that the excess charge is stored only at the surface of the capacitor plate, as indicated in Figure 9.1 by the yellow regions. The electrons are confined near the surface because they are attracted to the positive charges on the opposite plate. They would like to leap across the gap and join them, resulting in the unfortunate "lightning" mentioned above. The electrons in the rest of the metal see no net force because the attraction resulting from the positive charges on the surface of the other plate is exactly cancelled by the force resulting from the negative charges at the surface of their plate. The reverse holds true for the other plate. This "magic" balance occurs because the electrons in the metal are free to move anywhere in the metal. If there was a force in the metal inside the volume, the electrons would move toward the positive charges and away from the negative charges. They would continue to rearrange themselves and move around until moving does not provide them with any energy advantage. This results in the new charge being confined to very narrow regions at the surface of the metal. The remainder of the metal is electrically neutral and does not contribute to the potential. It adds only useless size and weight to the system. This is the fundamental problem of capacitors: almost all of their space is wasted.

Capacitive electrical energy storage has been around for hundreds of years without large improvements. In 1957, a technological revolution occurred, resulting in the development of supercapacitors. These systems no longer used flat plates. Instead they structured materials that store the positive and negative charges throughout a volume. One can understand the basic idea by considering two flat pieces of paper. Those pieces represent the surfaces on which the charge is stored. They take up a good deal of space, but rolling the pieces of paper reduces the space that was covered when the paper was flat, even though the length in the unrolled direction remains the same. In supercapacitors, the paper is crumpled. Crumpled paper clearly covers much less space than flat or rolled paper. This was the breakthrough of supercapacitors. They allow surface areas equal to the entire surface of the earth to be stored in capacitors that are less than 1 cubic centimeter in volume. Almost all capacitive storage for electronics applications now uses supercapacitors. This fabulous breakthrough was transforming, but it was not enough.

The fundamental problem with capacitors stems from their nature. Capacitors consist of pairs of conducting (metal) plates that are normally not charged. Work must be done to move charge from one plate to the other. That work can be stored very efficiently as electrical energy; however, the charge is stored only on the surface of the metal. The volume of the metal is entirely wasted. Furthermore, the surfaces can't be too close together; otherwise, "lightning" will transfer the charge from one metal to the other. Filling the space in between can improve this situation, but those materials result in some leakage of charge through the material, so the capacitors lose energy slowly with time. Improvements in capacitors thus involve increasing the ratio of surface area to volume, which increases the charge that can be stored in a given space. Nanotechnology offers great promise in that area. The second area is improving the filling material that separates the two types of charges so that the space can be small without leaking. At present, limitations in these two areas make the energy densities in capacitors much lower than those for batteries because batteries store charge throughout a volume rather than on a surface and the charge separation occurs as a natural function of the materials, rather than simply as a result of doing work, as will be discussed in detail below.

New capacitor breakthroughs are constantly being reported. For awhile, there were even stories about the commercialization of small cars based on ultracapacitors. Since the capacitors have infinite lifetimes and extremely rapid charging times, this would be extraordinary. Sadly, the promised devices have not yet appeared, but they are not impossible.

Batteries

Electrical potential energy storage also includes the chemical energy stored in batteries. Unlike capacitors that begin with two plates that like electrons equally well, batteries begin with two materials with different preferences for electrons. The difference is quantified by the electrode potential. Electrons will naturally move from a material with a lower electrode potential to a material with a higher electrode potential. The voltage produced by a particular battery terminal with a particular chemical composition is measured experimentally by measuring the voltage difference between a battery terminal with that chemical composition and a standard hydrogen terminal. Thus, by definition, the electrode potential difference for hydrogen is zero since electrons like both hydrogen terminals equally well. The fundamental problem with current batteries is that we have to live with the electrode potentials we are given by the structure of atoms. It is possible that some advance in nanomaterials will offer different energy structures, just as quantum wells have provided LEDs with energy gaps that differ from those occurring in natural crystals; however, the battery breakthrough has not yet come.

For illustration purposes, I will consider the Daniell cell, which is a very simple rechargeable battery made using zinc and copper. Zinc has an electrode potential value of -0.76 volts, whereas copper has a value of 0.34 volts. Thus, a battery made from the two should have a potential difference of approximately 1.1 volts, and electrons will flow from the zinc to the copper because that represents movement down an electrical potential hill. One of the reasons lithium ion batteries are so effective is that they provide the deepest energy valley per atom. Since lithium is a very light atom, batteries based on lithium provide some of the best energy-per-unit mass values.

It is desirable that the motion of the electrons be controlled by the user, so she can determine when energy will flow. In burning reactions, which also depend on spontaneous electrochemical reactions that gain energy by transferring electrons to states with lower energy, there is a potential energy barrier that prevents materials such as wood and gasoline from spontaneously combusting. In the case of batteries, the two chemical reactions would spontaneously to reduce their energy; therefore, battery makers keep the two sides of the battery physically separated so the reactants cannot come together spontaneously. When the user wants to draw electrical energy from the battery, she connects a wire between the two different sides of the battery. That physical connection allows the charge to flow between the two sides of the battery. This idea is illustrated in Figure 9.2. The boulder on the side of the hill represents electrons in the material that wants to give off electrons. They would move to the other material if they could, but their motion is blocked because there is no path that allows them to go downhill. Connecting a wire between the batteries opens up a path, and the charge then freely flows down the electrical hill from one side of the battery to the other. That downhill flow can do electrical work.

The challenge in batteries is to separate charges that want to come together and make sure that they can come together only by flowing through wires connected to a device, not by moving around inside the battery without going through the wire. The flow of charge through the connecting wire can light a light bulb or perform other work. The motion is spontaneous once the wire is connected because the motion of the electrons releases stored chemical energy.

The chemical reaction in a battery does not occur directly between the two metal electrodes, the chemistry in solution matters. The Daniell cell uses weak sulfuric acid as the electrolyte, which facilitates the transfer of charge between the two metals. The acid contains sulfate ions, SO_4, which in water have a net charge of -2, so chemists often designate the sufate ion as SO_4^{2-}, where the superscript indicates the charge of the ion.

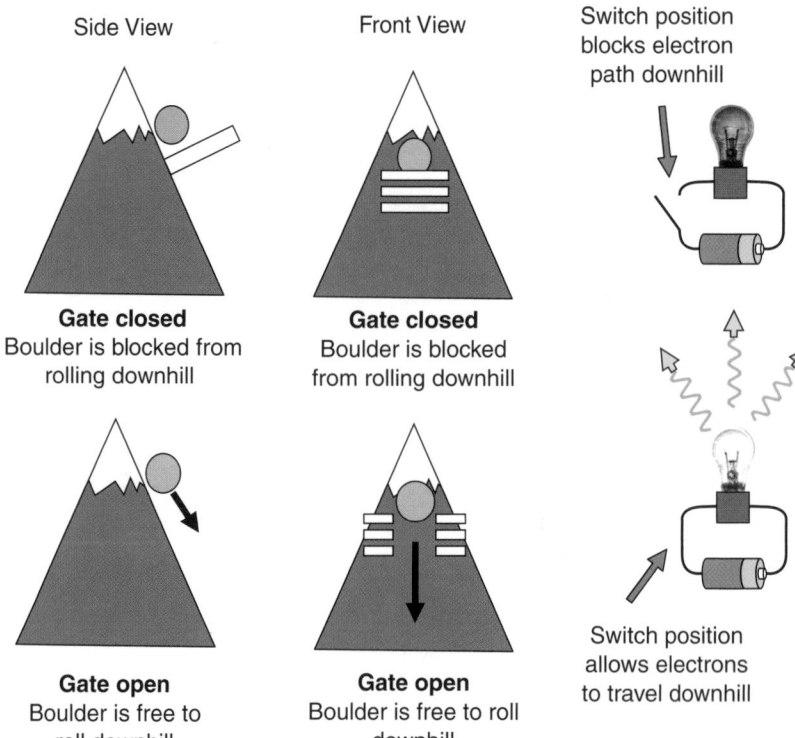

Figure 9.2 Illustration of how energy is stored in a battery. The left sides show the gravitational analogy, and the right column shows the battery system. When the two terminals of the battery are not connected, one side of the battery is at a much lower electrical potential than the other, but the charges do not have a path that would allow them to travel from the high-potential side to the low-potential side. Once the two terminals are connected, the wire offers a path, and the stored charge duly rolls down the electrical potential hill. That motion of the electrons can then do work, such as lighting a lightbulb.

In water, both zinc and copper can exist as ions with a net charge of +2, designated Zn^{2+} and Cu^{2+}, respectively. The zinc and copper ions combine with the sulfate ions to form neutral $ZnSO_4$, zinc sulfate, and $CuSO_4$, copper sulfate. Because of the higher electronegativity of copper, it is favorable to convert $CuSO_4$ to $ZnSO_4$.

When the cell is in its charged state, the copper side has lots of copper and sulfate ions free in solution, and the zinc side has very little free zinc.

The zinc spontaneously dissolves in the sulfuric acid, just as table salt dissolves in water. In the absence of a connection to the copper side, the solid zinc would reach equilibrium with the zinc in solution, just as adding a little bit of table salt to water dissolves the salt, but adding lots of table salt to water results in solid salt in salty water. Similarly, in the absence of a connection, the copper ions will coexist with the solid copper at equilibrium, again like solid salt in very salty water. In the presence of the sulfuric acid, it is energetically favorable for neutral zinc ions in the solid to give up two of the electrons to become Zn^{2+} in solution. The two electrons liberated by the formation of the Zn^{2+} ion from the electrically neutral zinc metal flow through the connecting wire from the zinc electrode to the copper electrode. At the surface of the copper electrode, the two free electrons that flowed through the wire from the zinc combine with the Cu^{2+} dissolved in the sulfuric acid, forming neutral solid copper metal at the surface of the copper electrode. Thus, the copper electrode grows in volume as more Cu^{2+} ions from solution are added to the solid copper electrode. Similarly, new zinc continues to be converted into Zn^{2+} ions in solution, resulting in a reduction in the volume of the solid zinc electrode. Thus, the battery can continue to work until it runs out of solid zinc; the energy is stored throughout the volume of the zinc metal. Thus, all of the volume of the copper and zinc contain charges that can be used to do work. In contrast, in capacitors, only the surfaces held charges that could do work. This is why batteries have better energy per unit volume than capacitors and why capacitor research tries to maximize the surface-area-per-unit volume.

The rate at which energy can be delivered or stored depends on the rate at which the chemical reactions between the electrodes and the solutions occur, whereas there was no rate-limiting factor in capacitors. Thus, high-energy storage applications normally use batteries, whereas high-power applications use capacitors because they require that a lot of energy be delivered in a short time.

The white tube in the zinc-copper battery illustration in Figure 9.3 is an additional connection between the two sides, which allows ions to

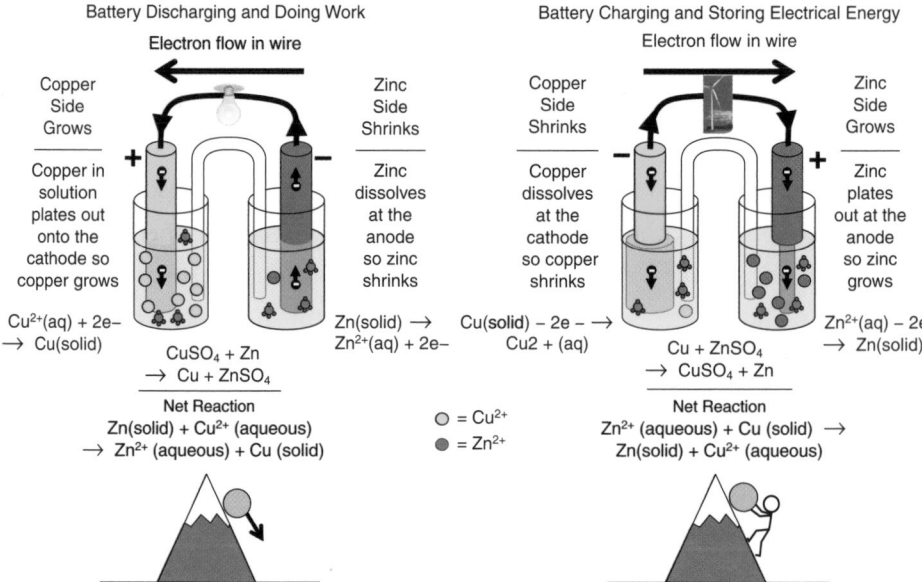

Figure 9.3 Illustration of a rechargeable battery system. The discharge of the stored chemical potential energy is illustrated in the upper panels. Here the electrons move along the wire from the zinc side to the copper side. The lower panels show mechanical work from a wind turbine being harnessed to create a potential difference between the zinc and the copper that causes the electrons to move along the wire from the copper side to the zinc side.

move. Without such a connection, the buildup of negative charge on the copper side, due to the arrival of the electrons, would stop the reaction. The moving ions keep both sides electrical neutral: the charge transferred through the wire by the electrons is canceled by the motion of the ions through the tube, which is usually filled with a salt that can provide ions but will not participate in the electrochemistry. Potassium chloride is a salt that can be used for this purpose. There will always be some leakage of zinc and copper ions through the connection. This leakage allows chemical reactions that do not require electron flow through the battery. Thus, this leakage represents an energy loss in the system. It is very important that the charge motion be dominated by discharge through wires, rather than leakage inside the battery.

Once the zinc is completely dissolved or the Cu^{2+} ions are removed from the solution, the battery is "dead." It cannot provide any more power because it is no longer favorable for electrons to flow through the wire. If the zinc electrode is not completely dissolved, the battery can be recharged by using an external source of electrical power, such as a windmill-driven generator, to do work to move electrons from the copper to the zinc through the wire. The excess electrons at the zinc electrode then interact with Zn^{2+} ions in solution, resulting in metallic zinc deposited on the electrode. At the same time, the copper electrode is shrinking because the metallic copper is losing electrons and becoming Cu^{2+} ions in solution. Thus, during recharging, the zinc electrode grows back, and the copper electrode loses its excess girth. A similar battery can be made at home using a lemon, a copper wire, and a zinc wire.

Real batteries do not consist of separate beakers of liquid, though lead acid batteries often do contain liquid, but the general principles in all batteries are the same. As illustrated in Figure 9.4, batteries consist of two materials with different electropotentials), so the electrons flow from the material with the lower electropotential to the material with the higher electropotential. That flow of electrons would normally build up a charge that would stop the battery from operating; thus, the change in charge due to the electrons has to be canceled by ions that move between the two sides of the cell and maintain charge neutrality on both sides. The electrolyte provides the ions that move from one side to the other. They are chosen to be ions that do not interfere with the chemistry between the electrodes. In Daniell cells, the negatively charged sulfate ions SO_4^2 can flow through the separator from the copper side to the zinc side in order to maintain charge neutrality. The electrolyte and the separator do not contribute directly to the chemical reaction providing energy, but they are required for the battery to function. Similarly, there must be a container to hold everything. The containers, the electrolyte, and the separator add size and weight to the battery without directly contributing to the energy. In contrast, all of the mass in methane or gasoline is used in the energy-generating chemical reactions, and they get to steal oxygen from

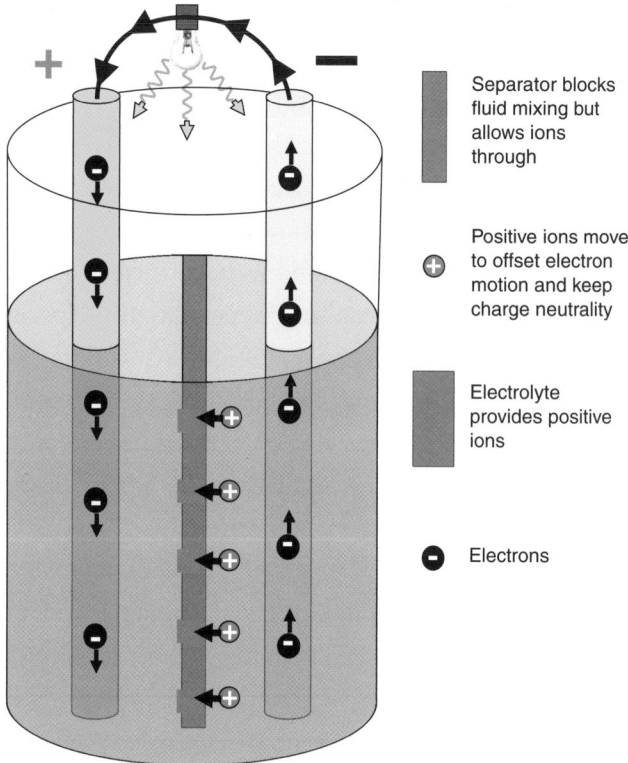

Separator blocks fluid mixing but allows ions through

Positive ions move to offset electron motion and keep charge neutrality

Electrolyte provides positive ions

Electrons

Figure 9.4 Illustration of a generalized battery showing the major features and flow of electrons when the two sides of the battery are connected by a wire, creating a flow of charge that can light a lightbulb or do mechanical work.

the air rather than carrying stored oxygen with them. All of these factors make the energy density of gasoline much more favorable than the energy density in batteries.

Another reason lithium ion batteries are special is that the mobile ions are lithium atoms. They move through the separator to balance the motion of the electrons, but they also participate in the chemical reactions. In fact, they determine the direction of the chemical reaction. In this case, both the positive and negative terminals are compounds that include lithium, but at the negative terminal lithium atoms will spontaneously come off the solid into solution, whereas at the positive terminal they spontaneously come out of solution and increase the volume of the solid. If an electrical potential is placed across the two terminals, the reaction direction is reversed and the battery recharges. One possible lithium ion battery chemistry is composed of an anode made from carbon and a cathode from a lithium cobalt oxide, where the electrolyte is a lithium salt in an organic solvent.

In capacitors, the potential difference between the electrodes depends only on how much work is done to move charges between them. The voltage difference can have any value up to the value where "lightning" occurs and the plates spontaneously discharge by moving across the space that separates the two plates. In contrast, the voltages on batteries are determined by the fundamental chemistry of the material since it depends on how much energy an electrons gains by being transferred from the less electronegative material to the more electronegative material. Thus, the voltage differences in batteries remain fairly stable until the battery is nearly depleted. In contrast, the voltages across capacitors constantly decrease as the stored energy is consumed; therefore, systems that use capacitors for energy storage must include additional circuit elements that manage the voltage decrease. Table 17 offers a comparison of the energy-storage properties of capacitors and batteries, indicating that batteries provide much higher energy densities but are less effective in providing power.

The shelf life of a battery is the length of time that the battery will remain charged if it is not used. The shelf life is determined by the self-discharge rate of the battery, which is the rate at which charge is transferred between the two halves of the battery if they are not connected. Lithium ion batteries have the highest energy density and the longest shelf life; however, they are very expensive and they typically function for approximately only three years whether or not they are actually used. This makes them well suited to applications in cell phones and laptops, where the devices themselves are likely to be replaced within

Table 17.   Comparison of the properties of capacitors and batteries, indicating that batteries provide larger energy storage, but capacitors provide more power and faster charging times.

Characteristic Quantity	Capacitor	Battery
Voltage vs. time	Voltage changes rapidly with use	Voltage nearly constant for most of its life
	→ Extra circuitry required for constant voltage applications	
Charge and discharge rates limit power that can be delivered	Charging and discharging rates can be infinitely fast, usually limited by other circuit elements	Charging and discharging rates are limited by chemical reaction rates
	→ High power delivery	→ Lower power delivery
Distribution of stored charge limits energy over volume	Useable charge stored on the surface only	Useable charge stored throughout volume
	→ Lower energy density	→ High energy density
Temperature dependence	Very little for most systems	Some chemistries depend strongly on temperature

three years. In contrast, NiMH batteries have lifetimes of up to ten years, which is probably why they were chose for the Prius. NiMH batteries have shorter shelf lives than lithium ion batteries, losing around 30 percent of their energy in a month, but that is not a large consideration in cars that are driven every day. NiMH batteries also have lower energy densities, so for a given battery volume the driving range will be smaller than the range for lithium ion batteries, which is probably why Tesla cars use the latter. Tesla believes its battery packs will last more than three years, offering an eight-year, 125,000-mile warranty on the Model S 60-kilowatt-hour battery and the same eight years but unlimited mileage on the 85-kilowatt-hour model.

Current batteries already exploit electronegativities that are near the limit, so choosing better atoms is unlikely unless there is a major materials science advance. Thus, increasing energy density is likely to involve clever strategies for storing more usable atoms without resulting in discharge within the battery itself. Nanotechnology breakthroughs in this area have been heralded for the last ten years, but so far they have not yet resulted in commercial products. At present, batteries are preferred in applications that require lots of energy storage, and capacitors are preferred in situations that require lots of power delivery. Various hybrid battery-capacitor systems are under investigation in an effort to exploit the best features of both systems.

Photosynthesis

In photosynthesis, plants take solar energy and use it create chemical energy that can later be consumed. A photosynthetic chemical reaction is listed below. It takes carbon dioxide and water and uses energy from light to produce carbohydrates and oxygen.

$$\text{Carbon dioxide} + \text{Water} + \text{Light} \rightarrow \text{Carbohydrates} + \text{Oxygen}$$

Most life depends on this reaction since it allows plants to obtain food, and animals obtain food by consuming plants or other animals that previously ate plants. Wood burning exploits this energy storage, as does biofuel-based ethanol. Polymer scientists are trying to develop artificial photosynthetic systems, but most research is related to making plants that more efficiently transfer solar energy directly to ethanol or another burnable fuel without requiring any additional processes.

KINETIC ENERGY–BASED STORAGE

Potential energy–based storage depend on an object remaining at some particular point on a potential hill until the energy is released by allowing the object to fall down the potential hill. The energy can remain stored forever as long as the particle does not move down the hill. In contrast, kinetic energy–based systems store energy in the motion of particles. No stored energy is left if the particles stop moving, which poses substantial problems for macroscopic objects; however, progress is being made, and, in some cases, kinetic energy–based systems offer substantial advantages over potential energy–based systems.

Flywheels

Once an object is moving, stopping the object requires energy. Superman stopping a runaway train illustrates this effect. Rotating wheels can be "spun up" and will continue to rotate for some time. Basically, they are giant tops that continue to rotate until friction removes all of their energy or until their energy is consumed by a user. Like hybrid cars, the flywheel system contains a device that can be used as either an electric motor or a generator. When energy is being stored, electricity rotates the motor, which spins up the flywheel. When electricity is being removed, the rotation of the flywheel makes the shaft in the generator spin, which results in electricity being produced by the generator. Using magnetic bearings rather than material bearings reduces friction and increases the time between required bearing maintenance.

Flywheel power storage systems have energy densities similar to batteries and higher power densities. Flywheels are mainly used to provide load leveling for large systems, such as an uninterruptible power supply (UPS) for data centers.[10] Though this storage is useful in such niche applications where power is the most important factor, wider adoption had not been expected; however, given that the costs of flywheel UPS are about $1,333 per one kilowatt for a system that can store 15 kilowatt hours of energy, some people argue that such systems would be suitable for residential power storage.[11] This suggestion is not unreasonable for rooftop solar systems that can generate 2 kilowatts per hour, since the

10. See beaconpower.com/ and www.power-thru.com/.

11. Chris Nelder "Turn Up the Juice: New Flywheel Raises Hopes for Energy Storage Breakthrough," April 10, 2013. *Scientific American*. Available at www.scientificamerican .com/article/new-flywheel-design/.

average household electricity consumption is approximately 1 kilowatt, and darkness rarely lasts more than 15 hours.

Superconductors

Electrons in superconductors move freely without losing any energy. Thus, once one invests the energy in them, they continue moving forever. This effect is exploited in the magnets used for magnetic resonance imaging. It is also used to provide backup energy storage for some electrical power plants. Superconductivity would be much more widely exploited if it did not require the superconductor to be kept at a very low temperature. For low-temperature superconductors, this is about 4 degrees kelvin or around −270°C. Even "high-temperature" superconductors required temperatures so cold that nitrogen becomes a liquid. Large-scale adoption of superconductor-based energy storage probably awaits the development of higher-temperature superconductors. Room-temperature semiconductors would revolutionize energy storage and also significantly increase energy efficiency. Their development continues to be eagerly awaited.

Compressed Air Storage

Compressed air storage is similar to pumped hydro, except high-pressure air is pumped into a reservoir.[12] Such systems are often used in conjunction with a natural gas.[13]

Hybrid trucks that use a combination of compressed air storage and conventional engines have been used in very large vehicles for many years. Recently, such systems have been adopted for smaller trucks. The United Parcel Service began testing delivery trucks that store energy in compressed

12. See www.lightsail.com/ and www.sustainx.com/technology-isothermal-caes.htm.
13. See www.dresser-rand.com/literature/general/85164-10-CAES. PDF page 4 discusses how the power is extracted. See also www.powersouth.com/mcintosh_power _plant/compressed_air_energy.

air tanks in 2007. Fuel efficiency in such trucks is expected to be around 50 percent better than the efficiency in conventional trucks.[14] Delivery and waste disposal trucks start and stop frequently, so great energy efficiencies can be obtained by recycling the same energy many times. In such applications, it is power storage rather than energy storage that is important because the energy from the brakes is immediately used to accelerate the car forward after the stop. The storage is not intended to provide energy for driving over long distances. For these applications, hydraulic systems are much more effective than battery systems because they transfer power much more effectively, though they cannot store as much energy as batteries.

Thermal Storage

Hot tea or coffee remains hot for while. It remains hot a great deal longer if it is kept in a sealed thermos because heat cannot readily enter or leave the thermos. Thus, it is clear that thermal energy can be stored in warm liquids. The amount of heat a fluid can store is proportional to its heat capacity. Water has a fairly high heat capacity, but the heat capacity of molten salts is even larger. Solar power stations that heat water or salt can use the hot liquid to generate electricity after the sun has gone down, as discussed in Appendix B. Such systems may make solar more economically feasible because backup fossil fuel burning would not be required. However, as was shown in Figure 4.2, the cost of solar thermal power is much larger than any other energy source other than offshore wind, so it is not clear that increasing the cost further by incorporating energy storage would be economically sensible.

New technology offers a form of thermal storage that involves transferring heat between two reservoirs, rather than simply storing it in one

14. See www.pressroom.ups.com/Fact+Sheets/UPS+First+In+Industry+to+Purchase +Hydraulic+Hybrid+Vehicles and www.sustainablebrands.com/news_and_views/articles /ups-deploys-40-hydraulic-hybrid-delivery-trucks.

as described above. The system is designed for grid-scale storage and depends on the same principles as heat pumps, which are discussed in Chapter 8.[15] The inventors claim such systems can "store and discharge electricity with a 'round-trip' efficiency of 72–80%. This compares with 74% for pumped-hydro, the way 99% of grid-scale storage happens at the moment." Better still, they claim they can do this at a levelized cost of $0.05 per kilowatt-hour.[16]

15. Monitor: Pumping heat," June 5, 2014. The *Economist*, from the print edition. Available at www.economist.com/news/technology-quarterly/21603184-reversible -heat-pump-promises-cheap-way-store-renewable-energy.
16. Ibid.

10

CONSEQUENCES OF CONSUMING ENERGY

In general, negative effects associated with energy use do not occur during the moment that the energy is used. Instead, the negative effects are results of the processes that are necessary to extract and process the primary energy source and to generate and deliver the energy as well as the waste products generated by using the energy. In this chapter, I will consider the consequences associated with energy use and try to separate out those that are unavoidable from those that could be changed using present science and technology.

Philosopher Arthur Schopenhauer once said "men are so apt to be deceived in their judgment as to what is good or bad for them."[1] Unfortunately, energy policy is an area that provides frequent illustrations of that aphorism. Overwhelming emotional responses to dramatic media images have often played a significant role in determining energy policy. Similarly, particularly well-publicized incidents with relatively small overall impact often play a greater role in policy decisions than such things as commonplace daily emissions whose cumulative effects can be vastly greater. Furthermore, recent economic theory has highlighted the extent

---

1. See ebooks.adelaide.edu.au/s/schopenhauer/arthur/counsels/chapter4.html, first paragraph of section 51.

to which even everyday consumer decisions are not made purely on the basis of rational comparison but are governed by irrational factors that can cause us to make disadvantageous decisions.

Even in an imaginary universe of perfectly rational actors, scientific facts about energy generation are not sufficient to determine what choices individuals, corporations, or governments may make. Such facts can provide a range of available choices and some projections of possible effects of those choices, but most of the factors governing energy choice are outside the realm of the physical sciences. Mathematically inclined people are drawn to establishing some measured quantity or metric that should be optimized for decisions. In practice, cost is often chosen as this metric. In such a system, the preferred option is simply the lowest-cost option. Of course, people often use different metrics. After all, if cost were the only metric, first-class seats on airlines would not exist.

Energy policy decisions are often made by people for whom one choice provides a clear, immediate, local gain, whereas the negative consequences of that choice are spread in space and time to people with absolutely no control over the decision. Such issues are tremendously difficult. Game theory research has shown that cooperation and altruism can confer significant evolutionary advantages, but traditionally humans have been more swayed by the needs and desires of those who are close to them than by those who are not. That idea of closeness can include physical, temporal, and genetic proximity. A famous philosophical exercise involves the choice between saving one's immediate family member at the cost of the lives of some enormous number of people that the decision maker has never met. Sadly, energy policy decisions often involve precisely that choice.

It is widely known that energy use practices in one country affect life in others. For example, fossil fuel burning in the United States is associated with acid rain in Canada. Of course, it is associated with acid rain in the United States as well, but at least affected U.S. citizens have a federal government structure through which they can attempt to influence decisions made elsewhere in the United States. Canadian citizens have far less power over U.S. political decisions, even though they may suffer significant consequences from those decisions. A more dramatic example of cross-border effects is provided by the Chernobyl nuclear disaster. The outside world discovered the disaster because workers at the Forsmark Nuclear Power Plant were setting off nuclear detectors at the plant even though the plant is located more than 600 miles from Chernobyl.[2] More recently, small amounts of emission from the Fukushima plant have been detected near Vancouver some 5,000 miles away from Fukushima.[3]

Policy decisions about energy are particularly complex because many people believe that some of the most important factors depend either on our understanding of the impact of highly improbable events or our appreciation of consequences that may not appear for more than a hundred years. Given the massive uncertainties associated with future consequences, this chapter will not attempt to provide any detailed assessment of the costs that energy-related decisions might have throughout the globe during the next one hundred years or the next 100,000 years. Rather, it will contain brief summaries of some issues that people have raised in connection with energy use as well as some short considerations of fundamental connections that cannot be altered. For example, it is a fundamental law that burning fossil fuels must generate carbon dioxide. That carbon dioxide may then be captured and sequestered, but as long as fossil fuels are burned in air, it is impossible to avoid creating the carbon dioxide in the first place. No "clean fossil fuel" can avoid that reality.

I divide possible consequences into three types: (1) consequences that occur before the energy is transformed to meet one of our needs, such as costs associated with mining and drilling; (2) the immediate direct

---

2. See sverigesradio.se/sida/artikel.aspx?programid=2054&artikel=4468603.

3. Becky Oskin and LiveScience, Radioactive Isotopes from Fukushima Meltdown Detected near Vancouver, *Scientific American*, February 25, 2014, www.scientificamerican .com/article/radioactive-isotopes-from-fukushima-meltdown-detected-near-vancouver/.

consequences of the transformation, such as the products created by burning; and (3) the more indirect consequences for the future associated with the changes in the world that we made in order to exploit the energy. These last issues are the most uncertain and the most difficult to address. Long-term consequences divide into two sorts: those associated with rare catastrophic events, such as large oil spills in oceans or nuclear power plant meltdowns; and those associated with small environmental changes that occur constantly at many locations all over the world. The first sort is more dramatic, but the second may have much more catastrophic long-term consequences because so many sources contribute to them for so long. Of course, all of these negative consequences can be reduced if we simply use less energy, which is why increases in energy conservation and increases in energy efficiency are so important. The long-term effects of energy choices are also often the consequences of which we are least conscious when making a decision, but as a society we can and do change.

In casual conversations, many people speak as if renewable energy use has no negative consequences. However, this is not the case, as I discuss in the following sections.

HYDROELECTRIC DAMS

Except for the very largest hydroelectric dams, the extraction of resources that is required in order to construct the dams represents such a minor impact on the mining and land management activities of the economies involved that examining possible consequences is not terribly worthwhile. For example, whatever environmental effects are associated with creating cement for hydroelectric dams, those effects are predominantly produced by users of cement for projects other than the building of dams. Similarly, emissions resulting from hydroelectric power plants are small.

For hydroelectric dams, the major effects of generation are the result of two factors: (1) changes related to the creation of the large reservoir behind the dam and (2) changes in the water flow of the river that was dammed. These two factors have can have environmental, social, cultural, and political effects, as I discuss below.

Ecological Changes

Small hydroelectric dams have existed for a long time. Such small dams do not dramatically change water flow or require the relocation of many people, but they can have a major impact on the life of animals that live in or near the river. Effects on fish migration, particularly salmon migration, have been a big issue in the United States. In some places, concerns have been so great that they have not only resulted in delays or cancellations in planned projects, they have resulted in the decommissioning of functioning dams.[4]

As time has passed, dams have become larger, as have the environmental impacts of the dams. When large-scale hydroelectric dams were first built, they were regarded as a wonderful clean-energy alternative to fossil fuels that would have no negative impact on the environment. The construction of Aswan Dam in Egypt provided the world with a laboratory that highlighted some of the effects of dam building that had not previously been appreciated; however, forming conclusions from the behavior of such complex systems where there are so many uncontrolled variables is extremely challenging. In *War and Peace,* Tolstoy pointed out that the fact that the train always comes when one's watch displays a certain time does not mean that one's watch displaying that time causes the train to come. The discussion of the environmental impact of Aswan Dam highlights the difficulty in disentangling changes that occurred as a consequence of just the dam from other environmental, commercial, and societal changes that would have occurred even if the dam had not been built.

For example, the annual flood of the Nile has been a feature of Egyptian life for thousands of years. It provided an annual deposition of sediment consisting of fine sand, silt, and clay. The millions of tons of annual sedi-

---

4. See www.ussdams.org/c_decom.html.

ment load of the Nile is now largely being deposited in Lake Nasser, which is the lake behind Aswan Dam.[5] Thus, the lake is slowly filling up. Some sedimentation removal project will eventually be required to keep the lake from becoming a mud bath. The sediments now filling Lake Nasser used to be deposited downstream, where they constantly restored the fertility of the soil without requiring application of chemical-based fertilizers. Soil fertility has significantly decreased since the building of the dam.[6] The use of chemical fertilizers in Egypt has vastly increased since the dam was built, but it is not clear how much of that increase is due to changes in flooding and how much is due to changes in farming practices that would have occurred even in the absence of the dam.[7] The "green revolution" was not caused by the building of the dam. Similarly, coastline erosion has been increasing. The coastline is now rapidly eroding rapidly,[8] but patterns of development and land use have also been changing. Finally, saline levels in farmland have increased since the building of the dam, but so has irrigation, which is known to increase salinity even in the absence of the changes in groundwater height that have been attributed to the dam.[9] Proponents of the dam claim that the cycles of flooding and drought that have long plagued the Nile Delta have been reduced by the presence of the dam, and that those positive effects combined with the benefits due to hydroelectricity make the overall effect of the dam highly positive.[10]

5. See www.history.com/this-day-in-history/aswan-high-dam-completed.

6. See www.pbs.org/wgbh/buildingbig/wonder/structure/aswan_high.html.

7. Asit K. Biswas and Cecilia Tortajada, "Impacts of the High Aswan Dam," chapter 17, figure 17.6, *thirdworldcenter.com,* www.thirdworldcentre.org/sprchap17.pdf; C. Tortajada, et al., editors. 2012. *Impacts of Large Dams: A Global Assessment* (Berlin Heidelberg: Springer-Verlag).

8. See ocean.tamu.edu/Quarterdeck/QD3.1/Elsayed/elsayed.html.

9. Asit K. Biswas and Cecilia Tortajada, "Impacts of the High Aswan Dam: 17.3 Impacts of the Dam," chapter 17, PDF p. 383, *thirdworldcenter.com,* www.thirdworld centre.org/sprchap17.pdf; C. Tortajada, et al., editors. 2012. *Impacts of Large Dams: A Global Assessment* (Berlin Heidelberg: Springer-Verlag).

10. Ibid.

China is renowned for its ability to focus on long-term goals and ignore small short-term issues that can strangle projects in other countries. Similarly, China allegedly has a much easier time making decisions that sacrifice the interests of a few for the interests of many. Finally, the Chinese government is not known for public discussion of negative consequences of its actions. However, now even the Communist Party is starting to express concerns about environmental problems spawned by Three Gorges Dam, including water pollution and landslides. Shockingly, the government's news agency published an article whose title on the agency's English-language Web site was "China Warns of Environmental 'Catastrophe' from Three Gorges Dam."[11]

Social and Cultural Costs

The building of Aswan Dam in Egypt generated an enormous amount of publicity in the 1960s for various reasons. The dam was funded by the Soviet Union, which also provided a great deal of technical assistance in the construction. The construction of the dam required the relocation of the Abu Simbel temple complex, which would otherwise have been completely underwater after the construction of the dam. Dramatic images of the dismantling and reconstruction of the temple often appeared in the press.. The complex was relocated in its entirety in 1968, on an artificial hill made from a domed structure, high above the Aswan High Dam reservoir.

The *New York Times* noted that, when discussing Three Gorges Dam, Chinese officials are often quite proud to note that it holds many world records, including world's biggest dam, biggest power plant, and biggest consumer of dirt, stone, concrete, and steel.[12] Chinese officials are less

---

11. See news.xinhuanet.com/english/2007-09/26/content_6796234.htm.

12. Jim Yardley, "Chinese Dam Projects Criticized for Their Human Costs," The *New York Times,* November 19, 2007, www.nytimes.com/2007/11/19/world/asia/19dam.html ?pagewanted=all&_r=0.

eager to admit that it also holds the world's record for displaced people, with a total of approximately one million,[13] representing around thirteen cities, 140 towns, and 1,350 villages. For comparison, approximately 100,000 people were relocated when Aswan Dam was built.[14] In China, efforts were made to maintain social cohesion, but such massive relocation will inevitably produce strains. Given that many of the people were relocated to areas just above the newly flooded area, population density in those areas suddenly showed large increases, which have put strains on social relations and caused environmental stress resulting from the increased exploitation of local resources, including soil erosion and deforestation.[15]

International Political Consequences

As the discussion of the effects of Aswan Dam highlights, building a dam changes water flow downstream. This is why building a dam in one country can produce protests in other countries downstream even if the dam is separated from any border by hundreds of miles. When building a dam, acting locally can transform globally. For example, it has been suggested that the construction of Aswan Dam has resulted in erosion of Israeli beaches because the sediment that used to wash up on the beaches is now filling Lake Nasser.[16] Although this is probably not the greatest source of political tension in the Middle East, it highlights the extent to which it is difficult to predict long-term environmental consequences of

13. See www.npr.org/templates/story/story.php?storyId=17784497 and www.mtholyoke.edu/~vanti20m/classweb/website/socialconsequences.html.

14. Thayer Scudder, "The Aswan High Dam Case," Section One: Saad el Aali—the Aswan High Dam, California Institute of Technology, www.hss.caltech.edu/~tzs/Aswan%20High%20Dam%20case, PDF p. 1.

15. Jim Yardley, "Chinese Dam Projects Criticized for Their Human Costs," *The New York Times*, November 19, 2007, www.nytimes.com/2007/11/19/world/asia/19dam.html?pagewanted=all&_r=0.

16. See www.israelweather.co.il/english/page2.asp?topic_id=72&topic2_id=123&sub_topic_id=1.

projects that produce significant change in water flow. Even areas with lower levels of endemic political tension can have controversies sparked by dam building. For example, Laos has proposed building a hydroelectric dam on the Mekong River, but Cambodia and Vietnam are vigorously protesting against the construction of the dam. Any large effort to increase hydroelectric power by damming rivers that flow between countries is likely to be met with significant negative reactions from countries downstream from the project. The cost of resolving such issues must be included in considerations of new hydroelectric projects.

WIND POWER

As with hydroelectric dams, the effect of extracting the materials required to build wind power systems is dwarfed by other manufacturing projects that use the same materials, and the impact of such material extraction will probably not significantly influence policy decisions concerning wind power. Large-scale wind power projects are much more recent than large-scale hydroelectric projects; therefore, not enough time has elapsed for any long-term consequences to become clear. Wind farms do not emit carbon. Land use competition is less of a problem than it is for biofuel farming, for example, because land used for wind power generation can also still be used for farming. For wind farms in inhabited areas, particularly coveted coastline locations, ruined views are a significant issue. So far, people have expressed concern about effects associated with audible and inaudible vibrations, but a definitive study has not yet emerged. Bird deaths resulting from collisions with turbines have also been mentioned as an issue. Wildlife Society reports 573,000 bird deaths due to wind farms in 2012.[17] According to a 2002 study by the U.S. Fish and Wildlife Service, the bird population of the United States varies between

17. K. Shawn Smallwood, "Comparing bird and bat fatality-rate estimates among North American wind-energy projects," *Wildlife Society Bulletin,* March 2013, vol. 37, issue 1, pp. 19–33. The abstract is free and available at onlinelibrary.wiley.com/doi

10 and 20 billion, depending on the season.[18] That 2002 study makes the following estimates of annual bird deaths in the United States due to (1) window strikes: 100 million to 1 billion; (2) communication towers: 40 to 50 million; (3) collisions with cars: 60 million (4) high-power transmission lines: more than 170 million; and (5) cats, both domestic and feral: exceeding 50 million. A 2013 study published in the peer-reviewed journal *Nature Communications* indicates that those earlier estimates of annual bird deaths due to cats is too low, with more accurate estimates ranging between 1.3 and 4.0 billion.[19] A 2012 article in *Nature* reviewed the 2002 U.S. Fish and Wildlife service data, which they summarize nicely in a figure, emphasized that large, endangered, predatory bird species may be more endangered by wind turbines than domestic cats,[20] suggesting that care must be taken when simply comparing raw numbers. A final and more subtle issue is the possible effect of large-scale wind farms on weather and climate given that they significantly alter wind speed and direction. More detailed long-term monitoring will be required to accurately characterize such large-scale, long-term effects.

SOLAR POWER

Some exotic solar systems use materials that can have serious environmental impacts when they are mined or disposed of, but at present most solar photovoltaic systems use the same technology as computer chips.

/10.1002/wsb.260/abstract; See www.bloomberg.com/news/2013-12-06/u-s-eases
-turbine-bird-death-rule-as-cats-kill-millions.html.

18. See www.fws.gov/birds/mortality-fact-sheet.pdf.

19. Scott R. Loss, Tom Will and Peter P. Marra, "Corrigendum: The impact of free-ranging domestic cats on wildlife of the United States," *Nature Communications, December 12, 2013*, vol. 4, article 2961 doi:10.1038/ncomms3961. The abstract is free and the link to the corrected version is www.nature.com/ncomms/2013/131212/ncomms 3961/full/ncomms3961.html.

20. Meera Subramanian, "The trouble with turbines: An ill wind," *Nature.com*, June 20 2012, vol. 486 p. 311.

Once again, then, the effect of fabricating solar panels is dwarfed by the effects of fabricating semiconductors, so solar panel fabrication issues are unlikely to have a significant impact on decisions.

When in use, solar panels generate no emissions. Land use competition is not a problem for the two favored locations: deserts and roofs of dwellings. Neither location affects crop yields or housing. However, it is unclear how delicate desert habitats might be affected if their surface is covered in solar panels that remove all of the light and divert rainfall.

GEOTHERMAL POWER

In geothermal systems, water heated by the inside of the earth is used to provide climate control or mechanical work. In some places, sufficiently hot water emerges naturally, but in most places, a well is drilled and water is forced into the earth. As long as the water is used in a closed loop system, emissions associated with heating the water are small. Some people have suggested that seismic activity is associated with the local cooling of normally hot areas of the earth, but seismic activity is tremendously unpredictable.

CONSEQUENCES OF COMBUSTION

Just as there is universal agreement that pursuing renewable energy is good, there is nearly universal agreement that combustion-based energy generation is bad; however, fossil fuel burning is increasing rapidly, particularly in the developing world. At present, in most places, it represents the most flexible and inexpensive way to generate electricity. Similarly, it dominates vehicular transport and is likely to do so for the foreseeable future. Thus, like cigarette smoking, everyone knows that it is bad, but quitting is extremely difficult. Similarly, as with cigarette smoking, the immediate consequences are immediately positive for the user, whereas the negative consequences are usually delayed by tens of years. Thus, quitting smoking involves giving up an immediate pleasure to avoid negative future consequences whose impact is unclear. This is why governments have been considering putting pictures of diseased lungs on the outside of cigarette packages: they are supposed to remind the user of how bad that as yet unapparent future consequence may really be.[21]

Just as "low tar and nicotine" cigarettes are offered to try to reduce negative impacts from smoking for people who are unable to quit, significant effort is being expended to decrease the negative consequences associated with fossil fuel burning because the world is unlikely to quit burning fossil fuels in the near future.

The environmental impact of extracting fossil fuels for burning depends enormously on the fossil fuel being extracted, the location of the mine or well, and the particular extraction technique being used. The effects of strip mining coal are quite different from the effects of underground coal mining. Similarly, deep-ocean oil rigs have very different associated risks than land-based wells.

EXTRACTION CONSEQUENCES

Assigning "true costs" to energy consumption is challenging. It is also a crucial issue in evaluations that compare renewable energy to fossil fuel energy, because hidden costs and subsidies may create large differences between the direct costs paid by consumers purchasing energy derived from a particular fuel source and the global costs associated with the extraction and consumption of the energy; however, since the 1970's, the United States has been trying to include some environmental costs of extraction into the price of energy.[22]

21. See www.telegraph.co.uk/health/3085312/Pictures-of-diseased-lungs-on -cigarette-packets.html.
22. See www.osmre.gov/about.shtm.

Coal

For underground mining of coal, the effect of mining on the miners' health has traditionally been substantial. "Black lung disease" used to be fairly common among coal miners. Coal mining was an unpleasant life, though the closing of underground mines was widely protested by the miners themselves. Margaret Thatcher's political career received a significant boost from her successful defeat of a coal miners' strike triggered by threatened mine closings. Modern underground mining is less hazardous, but mining remains one of the most dangerous jobs in the United States.

Various surface mining techniques exist with names like strip mining and mountain top removal. Surface mining of coal completely alters the upper layer of the earth. Prior to the late twentieth century, land reclamation projects were uneven at best. In some cases, the mined regions were simply left exactly as they were the day the shovels stopped extracting coal.

The rise in environmental concern in the 1960s and 1970s resulted in political pressure to establish some guidelines for the fate of mined land. Powerful ads showed an apple being peeled, followed by an attempt to glue the skin back using white glue. Obviously, the attempt was unsuccessful. Once peeled, the apple could never be returned to its original state. The ad argued that strip mining was like apple peeling and that no remediation effort could ever be successful.

The Surface Mining Control and Reclamation Act of 1977 (SMCRA) established guidelines for existing and future mines as well as a trust fund to finance the reclamation of abandoned mines,[23] though significant regulatory activity still occurs at the state level. Funding for the reclamation of mines that have already been abandoned is provided by a tax. This tax provides an example of costs associated with energy use being included in the price of the energy. The proceeds from this tax are put

---

23. See www.osmre.gov/about.shtm.

into the Abandoned Mine Reclamation Fund, which is divided between state and federal programs. As of December 15, 2011, the Office of Surface Mining Reclamation and Enforcement (OSMRE) had provided more than $7.2 billion to reclaim more than 295,000 acres of hazardous high-priority abandoned mine sites and for other purposes for abandoned mine lands that have been reclaimed through the OSMRE fund since 1977.[24]

Not only can the mines themselves produce deleterious effects, but the processing following the mining can also produce hazards. In January 2014, there was an accidental release of chemicals used to process coal. That release in West Virginia caused widespread water contamination, which rendered local water undrinkable. The effects of the spill extended over nine counties.[25]

Mining not only affects the land that is mined and its surrounding ecology, but it can also release gases that alter the global atmosphere. For example, methane, a potent greenhouse gas that is trapped along with the coal, is often vented during these processes to increase safety. For the most part, however, the negative effects of coal mining are largely confined to the coal mining regions themselves. In contrast, the effects of emissions due to burning coal are global.

Natural Gas

In traditional natural gas wells, some methane is emitted during the extraction process, which also requires some fossil fuel burning. Similarly, methane can be emitted as the result of leaks and losses during transportation. The big issues concerning natural gas extraction are related to the new processes that have dramatically increased extraction rates and reserves. A 2013 study suggested that methane emission may be two

24. See www.doi.gov/news/pressreleases/Secretary-Salazar-and-Director-Pizarchik -Announce-485-Million-in-Grants-to-States-and-Tribes-to-Clean-Up-Abandoned -Coal-Mines.cfm.

25. See edition.cnn.com/2014/01/09/us/west-virginia-contaminated-water/.

times higher than estimated by the EPA.[26] Although some of this increase in methane emissions may be partly due to livestock emission, the result also indicates that emissions related to oil and gas extraction could be five times higher, based on methane measurements in regions with significant extraction activity.[27] It should be noted that the measurements showed only that there have been increases in methane emissions but did not directly show that the extraction activity produced those emission increases. The most certain outcome of the study is that more study is needed. Fortunately, methane emissions dissipate much faster than carbon dioxide emissions, so reductions in methane emissions will result in decreases in atmospheric methane within two decades.

It has been suggested that during "fracking" the chemicals used to extract natural gas may contaminate groundwater. As a dramatic counterresponse, in August of 2011 Halliburton Co., CEO Dave Lesar asked another unnamed Haliburton executive to drink Clean Stim,[28] a fluid used in "fracking," to demonstrate that it was safe.[29] The executive sipped and survived, but that does not prove much except that it is not immediately fatal in whatever dose he consumed it. It is also impossible to assess long-term effects of adding large quantities of chemical-laced water to select regions of the earth since this activity has only recently begun. Some recent work suggests that earthquake activity may be increased by fracking,[30] and that even the disposal of waste water from fracking may

26. Scott Miller, et al. "Anthropogenic emissions of methane in the United States," October 18, 2013, Proceedings of the National Academy of Sciences of the United States of America, doi: 10.1073/pnas.1314392110, www.pnas.org/content/early/2013/11/20/1314392110.full.pdf+html?with-ds=yes.

27. Andrew Revkin, "New Study Finds U.S. Has Greatly Underestimated Methane Emissions," *New York Times*, November 25, 2013, dotearth.blogs.nytimes.com/2013/11/25/new-study-finds-u-s-has-underestimated-methane-levels-in-the-atmosphere/.

28. "CleanStim® Hydraulic Fracturing Fluid System," *Haliburton.com*, www.halliburton.com/en-US/ps/stimulation/fracturing/cleanstim-hydraulic-fracturing-fluid-system.page.

29. See www.cbsnews.com/news/can-you-drink-fracking-fluid-one-gas-exec-did/.

30. Bryan Walsh, "The Seismic Link Between Fracking and Earthquakes," *Time* magazine, May 1, 2014, available at time.com/84225/fracking-and-earthquake-link/.

pose an earthquake threat;[31] however, more long term work would be required to more accurately assess the relationship between earthquakes and fracking.

Oil

Traditional land-based oil wells have limited environmental effects. The oil well fires deliberately set by Iraqi soldiers leaving Kuwait burned uncontrolled for months, resulting in dramatic images. The fires were the subject of a 1992 IMAX documentary film, *Fires of Kuwait,* which was nominated for an Academy Award.[32] The film includes footage of a Hungarian team of fire fighters using their jet turbine extinguisher. The Kuwaiti oil fires are also featured in Werner Herzog's 1992 film, *Lessons of Darkness.* The fires reportedly increased respiratory distress among Kuwaitis, but predictions of global weather disasters proved false.[33] More mundanely, but with more widespread effect, oil wells and oil collection equipment are a source of emissions of methane, and the large engines that are used in the oil drilling process burn natural gas or diesel that also produces emissions.

Oil transport from wells to consumers also requires fossil fuel burning that generates emissions, but more publicity is received from tanker accidents that occur during oil transport. The most famous tanker accident is the *Exxon Valdez* oil spill, which occurred in Prince William Sound, Alaska, on March 24, 1989. The *Exxon Valdez* was bound for Long Beach,

---

See stateimpact.npr.org/texas/tag/earthquake/; www.nbcnews.com/science/science-news/confirmed-fracking-practices-blame-ohio-earthquakes-f8C11073601; Rick Jervis, "Fracking wells possible culprit of Texas earthquakes," *USA Today,* June 1, 2014, www.usatoday.com/story/news/nation/2014/06/01/earthquakes-texas-fracking-wells/9765659/; Jim Efstathiou Jr., "Oklahoma Swamped by Surge in Earthquakes Near Fracking," *Bloomberg.com,* April 8, 2014, www.bloomberg.com/news/2014-04-07/oklahoma-swamped-by-surge-in-earthquakes-near-fracking.html.

31. See www.nbcnews.com/science/science-news/confirmed-fracking-practices-blame-ohio-earthquakes-f8C11073601.

32. See www.imdb.com/title/tt0104275/.

33. See en.wikipedia.org/wiki/Kuwaiti_oil_fires.

California, carrying oil originally extracted at the Prudhoe Bay oil field when it struck Prince William Sound's Bligh Reef and spilled 260,000 to 750,000 barrels (41,000 to 119,000 cubic meters) of crude oil. Prince William Sound's remote location, accessible only by helicopter, plane, and boat, made containment and cleanup particularly difficult. The spilled oil eventually covered 1,300 miles (2,100 kilometers) of coastline and 11,000 square miles (28,000 square kilometers) of ocean. After the *Exxon Valdez* oil spill, the U.S. government required that all new oil tankers built for use between U.S. ports be equipped with a full double hull. Though double hulls protect against low-impact collisions, higher-impact collisions will still penetrate both hulls, resulting in a spill.

The advent of deep sea–based oil drilling has resulted in a vast increase in the possible negative environmental consequences of oil drilling. The Santa Barbara oil spill that occurred in January and February 1969 enormously increased awareness of the possible effects of oil drilling that occurs in oceans. Pictures of injured sea birds produced an outpouring of emotion that colored U.S. policy toward offshore drilling for a generation.

Of course, the Santa Barbara oil spill was dwarfed by the 2010 *Deepwater Horizon* oil spill in the Gulf of Mexico. Like many accidents, it was not the result of a single failure but the consequence of a series of failures. In January 2011, the White House Oil Spill Commission released its final report on the causes of the oil spill. It blamed BP and its partners for making a series of cost-cutting decisions. It also concluded that the spill was not an isolated incident caused by "rogue industry or government officials" but that "The root causes are systemic and, absent significant reform in both industry practices and government policies, might well recur."[34] After its own internal probe, BP admitted that it had made mistakes that

---

34. Deep Water: The Gulf Oil Disaster and the Future of Offshore Drilling—Report to the President. National Commission on the BP *Deepwater Horizon* Oil Spill and Offshore Drilling, January 11, 2011, p. 122. www.gpo.gov/fdsys/pkg/GPO-OILCOM MISSION/content-detail.html.

led to the Gulf of Mexico oil spill. In June 2010, BP set up a $20 billion fund to compensate victims of the oil spill.[35] The policy debate following the spill highlighted some of the tensions surrounding energy decisions. Some people correctly argued that no well can ever be safe. Others argued quite correctly that shutting down drilling would increase unemployment in oil industry workers. Still others argued rather less factually that increased offshore drilling would allow the United States to achieve independence from imported oil, which is not consistent with our current understanding of oil reserves; however, fabulous unforeseen discoveries cannot be ruled out altogether. Furthermore, improved techniques for extracting oil from rock may increase domestic oil production more than deep-water drilling. Issues concerning the circumstances under which deep-water oil drilling should be allowed are likely to remain for generations.

CONSEQUENCES OF BURNING

Burning inevitably releases chemicals that were previously bound inside the fuel. If the chemicals released in burning are released in the atmosphere, they can have significant local and global effects now and into the future. One benefit of converting to wind or solar power is a reduction in these emissions; however, as discussed previously, though biofuels may be carbon neutral, burning them can generate emissions of some pollutants that greatly exceed the emissions from burning natural gas or gasoline.[36]

Pollutants can also be classified as either primary pollutants or secondary pollutants. A primary pollutant is one that is emitted into the atmosphere directly from the source of the pollutant and retains the same chemical form. An example of a primary pollutant is the ash produced

---

35. See en.wikipedia.org/wiki/Deepwater_Horizon_oil_spill.

36. For example, ethanol emits more ozone than the equivalent amount of gasoline does, and woodstoves emit more particles than do natural gas furnaces.

by the burning of solid waste. A secondary pollutant is one that is formed by atmospheric reactions of precursor or primary emissions. Secondary pollutants undergo a chemical change once they reach the atmosphere. An example of a secondary pollutant is ozone created from organic vapors given off at a gasoline station. The organic vapors react with sunlight in the atmosphere to produce the ozone, the primary component of smog. Control of secondary pollutants is generally more problematic than that of primary pollutants, because mitigation of secondary pollutants requires the identification of the precursor compounds and their sources as well as an understanding of the specific chemical reactions that result in the formation of the secondary pollutants.[37]

The increased environmental awareness that occurred in the late 1960s led to the establishment of the Environmental Protection Agency. The EPA was proposed by President Richard Nixon and began operation on December 2, 1970. The mission of the EPA was to protect human health and the environment by writing and enforcing regulations based on laws passed by Congress. The Clean Air Act requires EPA to set National Ambient Air Quality Standards for six common air pollutants: lead, particle pollution (often referred to as particulate matter), nitrogen oxides, ground-level ozone, carbon monoxide, and sulfur oxides.

Of the six pollutants, particle pollution and ground-level ozone are the most widespread health threats. The EPA calls these pollutants "criteria" air pollutants because it regulates them by developing human health-based and/or environmentally based criteria for setting permissible levels. The set of limits based on human health is called primary standards. Another set of limits intended to prevent environmental and property damage is called secondary standards. Efforts are in progress to allow the EPA to set emission standards for carbon dioxide as well because of its role in global climate change, but at present this issue has not been resolved. The following presents a brief discussion of the six major pollutants and their effects according to statements made by the EPA.

37. See www.epa.gov/apti/course422/ap2.html.

THE SIX MAJOR POLLUTANTS
MONITORED BY THE EPA

The United States Environmental Protection Agency (EPA) is charged with monitoring six major pollutants. The following is a summary of EPA statements about these pollutants, which are described in detail at www .epa.gov/air/urbanair/. Though energy consumption plays a role in the emission of all of these pollutants, in some cases it is not the dominant source of the pollution.

Lead

The dramatic decrease in lead emissions from motor vehicle exhaust demonstrates how effectively changes in policy can alter negative effects associated with energy use as long as an alternative is readily available. Tetraethyl lead, abbreviated TEL $(CH_3CH_2)_4Pb$, was added to gasoline as an inexpensive additive beginning in the 1920s to make engines run more smoothly and to prevent exhaust valve and seal wear, but it is toxic to humans and interacts poorly with catalytic converters.[38] As of January 1, 1996, the Clean Air Act banned the sale of leaded fuel for use in on-road vehicles.[39] Possession and use of leaded gasoline in a regular on-road vehicle now carries a maximum $10,000 fine in the United States. It is still used as an additive in aviation fuel for piston engine–powered aircraft and off-road vehicles; however, lead emissions from fuel combustion has been drastically reduced.[40] In 1975, 82 percent of the lead emissions in the United States was due to on-road vehicles, but by 1995, the contribution was negligible, though off-road vehicles continue to contribute around 18 percent of the total lead emissions, and metal processing

38. See en.wikipedia.org/wiki/Tetraethyllead.

39. See yosemite.epa.gov/R10/airpage.nsf/webpage/Leaded+Gas+Phaseout.

40. "Between 1970 and 2005, estimated nationwide lead emissions decreased by 99 percent (220,000 tons), mostly due to elimination of lead from gasoline for on-road vehicles" according to the EPA as stated at cfpub.epa.gov/eroe/index.cfm?fuseaction =detail.viewInd&lv=list.listbyalpha&r=216603&subtop=341.

was the dominant source accounting for about 50 percent of the total.[41] In 2011, off-road vehicles accounted for approximately 60 percent of lead emissions, with industrial processes providing almost 30 percent.[42] In 2012, the overall average lead emission in the United States was only approximately 10 percent of what it was in 1980.[43]

Despite dramatic improvements, approximately one-third of the population of the United States experiences air quality concentrations above recommended levels. Thus, there is significant motivation to make changes in energy use that lower the emissions of these pollutants.[44]

Particulate Matter

Particulate matter (PM) is the general term for a heterogeneous mixture of solid particles and liquid droplets found in the air, including dust, dirt, soot, smoke, and liquid droplets. Particles can be suspended in the air for long periods of time. Some particles are large or dark enough to be seen as soot or smoke. Others are so small that individually they can be detected only with an electron microscope. PM can be a primary or secondary pollutant. "Primary" particles, such as dust or black carbon (soot), are directly emitted into the air. They come from a variety of sources, such as cars, trucks, buses, factories, construction sites, tilled fields, unpaved roads, stone crushing, and burning of wood. "Secondary" particles are formed in the air from the chemical change of primary gaseous emis-

41. Numerical data provided in a spreadsheet available at oaspub.epa.gov/eims/eims
.roereport.getfile?p_download_id=11210. A graphical version of the data is shown in
exhibit 2-5 in the EPA's 2008 Report on the Environment, May 2008. Document number
EPA/600/R-07/045F, pp. 2–9. www.epa.gov/roe/docs/roe_final/EPAROE_FINAL_2008
.PDF.

42. U.S. EPA Air emissions sources Lead (NEI2011 v1 GPR). www.epa.gov/cgi-bin
/broker?_service=data&_debug=0&_program=dataprog.national_1.sas&polchoice=Pb.

43. See www.epa.gov/airtrends/lead.html. More dramatical levels of the 10th
percentile have dropped by a factor of 30.

44. Our Nation's Air: Status and Trends through 2010. U.S. Environmental Protection Agency, Office of Air Quality Planning and Standards, February 2012. Report #
EPA-454/R-12 001, pp. 6–7. www.epa.gov/airtrends/2011/.

sions. They are indirectly formed when gases from burning fuels react with sunlight and water vapor. These can result from fuel combustion in motor vehicles, at power plants, and in other industrial processes. PM2.5 describes the "fine" particles that are less than or equal to 2.5 micrometers in diameter. PM10 refers to all particles less than or equal to 10 micrometers in diameter.[45]

Nitrogen Oxides

Nitrogen oxides represent a family of molecules that include both oxygen and nitrogen. The general abbreviation NO_x refers to all of the different forms. In the expression N represents nitrogen, O represents oxygen, and x is an integer. NO_2 is the emission that is officially controlled by the EPA. When any of these oxides dissolve in water and decompose, they form nitric acid (HNO_3) or nitrous acid (HNO_2), which contribute to pollution effects that have been observed and attributed to acid rain. At present, transportation is the dominant source of NO_x emission, accounting for 62 percent of the total, with electric power generation coming in second at 31 percent.

Sulfur Oxides

Sulfur oxides (SO_x) are colorless gases formed by burning sulfur. SO_x gases are formed when fuel containing sulfur, such as coal and oil, is burned and when gasoline is extracted from oil or metals are extracted from ore. Sulfur dioxide (SO_2) is the criteria pollutant that is the indicator of sulfur oxide concentrations in the ambient air. SO_2 dissolves in water vapor to form acid, which is also associated with acid rain. Over 65 percent of the SO_2 released to the air, or more than 13 million tons per year, comes from electric utilities, especially those that burn coal. Other sources of SO_2 are industrial facilities that derive their products from raw materials like metallic ore, coal, and crude oil or that burn coal or oil to produce

45. See www.epa.gov/apti/course422/ap5.html.

process heat. Examples are petroleum refineries, cement manufacturing, and metal processing facilities. Also, locomotives, large ships, and some nonroad diesel equipment currently burn high sulfur fuel and release SO_2 emissions.

Carbon Monoxide

Carbon monoxide (CO) is a colorless, odorless gas formed when carbon in fuel is not burned completely. Motor vehicle exhaust contributes about 60 percent of all CO emissions nationwide.[46] Other sources of CO emissions include industrial processes (such as metals processing and chemical manufacturing), residential wood burning, and natural sources, such as forest fires. Woodstoves, gas stoves, cigarette smoke, and unvented gas and kerosene space heaters are sources of CO indoors. Higher levels of CO generally occur in areas with heavy traffic congestion. In cities, 95 percent of all CO emissions may come from motor vehicle exhaust; however, on-road vehicles account for only 56 percent of total emissions. Though there are fewer nonroad vehicles than road vehicles, nonroad vehicles are such heavy emitters that they account for 22 percent of all CO emission. CO levels are usually highest during the winter, when CO automotive emissions are greater and nighttime inversion conditions are more frequent. In inversion conditions, the air pollution becomes trapped near the ground beneath a layer of warm air.[47]

Ozone

Unlike the other five pollutants monitored by the EPA, ozone is not emitted directly into the air by combustion. Ozone is a secondary pollutant that at ground level is created by a chemical reaction between

46. Latest Finding on National Air Quality 2002 Status and Trends. U.S. Environmental Protection Agency, Office of Air Quality Planning and Standards Emissions, Monitoring and Analysis Division, August 2003. Report #EPA 454/K-03-001, p. 16.

47. See www.epa.gov/apti/course422/ap5.html.

oxides of nitrogen (NO_x) and volatile organic compounds (VOCs) in the presence of heat and sunlight. Ozone is a triatomic molecule consisting of three oxygen atoms. It is a colorless compound that has an odor that people usually associate with electrical sparks. The concentration of ozone in a given locality is influenced by many factors, including the concentration of NO_2 and VOCs in the area, the intensity of the sunlight, and the local weather conditions. Ozone and the chemicals that react to form it can be carried hundreds of miles from their origins, causing air pollution over wide regions. Ozone contributes to "smog" or haze. Ozone pollution can be greatly reduced by reducing the emission of NO_2, which results from fossil fuel combustion, particularly gasoline engines.

Ozone is a powerful oxidant, which means that it readily interacts with other chemicals that can lower their energy by allowing their electrons to be more closely associated with the oxygen. As a result, ground-level oxygen causes ozone to damage mucus and respiratory tissues in animals and also tissues in plants above concentrations of about one hundred parts per billion. This makes ozone a potent respiratory hazard and pollutant near ground level. People with lung disease, children, and older adults are particularly at risk. Ironically, when the same ozone is present high above the earth in the stratosphere, the very same ozone molecule is beneficial because it prevents damaging ultraviolet light from reaching the earth's surface. The levels of undesired ground level ozone can be reduced by reducing the nitrogen oxides emissions that cause the ozone. NO_x emissions are dominated by transportation, but coal-based electric power generation also makes a significant contribution. Table 18 provides a summary of the health effects of the six major pollutants.[48]

48. The original version of Table 18 is available at www.epa.gov/airtrends/2011/report /airpollution.pdf. It is the first table in the document.

Table 18. The sources and adverse effects of the six major pollutants regulated by the EPA.

Pollutant	Sources	Health effects
Ozone (O_3)	Secondary pollutant typically formed by chemical reaction of volatile organic compounds and NO_x in the presence of sunlight.	Decreases lung function and causes respiratory symptoms, such as coughing and shortness of breath; aggravates asthma and other lung diseases, leading to increased medication use, hospital admissions, emergency department (ED) visits, and premature mortality.
Particulate matter (PM)	Emitted or formed through chemical reactions; fuel combustion (e.g., burning coal, wood, diesel); industrial processes; agriculture (plowing, field burning); and unpaved roads.	Short-term exposures can aggravate heart or lung diseases, leading to respiratory symptoms, increased medication use, hospital admissions, ED visits, and premature mortality; long-term exposures can lead to the development of heart or lung disease and premature mortality.
Lead	Smelters (metal refineries) and other metal industries; combustion of leaded gasoline in piston engine aircraft; waste incinerators; and battery manufacturing.	Damages the developing nervous system, resulting in IQ loss and impacts on learning, memory, and behavior in children. Cardiovascular and renal effects in adults and early effects related to anemia.
Oxides of nitrogen (NO_x)	Fuel combustion (e.g., electric utilities, industrial boilers, and vehicles) and wood burning.	Aggravates lung diseases, leading to respiratory symptoms, hospital admissions, and ED visits; increased susceptibility to respiratory infection.
Carbon monoxide (CO)	Fuel combustion (especially vehicles).	Reduces the amount of oxygen reaching the body's organs and tissues; aggravates heart disease, resulting in chest pain and other symptoms, leading to hospital admissions and ED visits.
Sulfur dioxide (SO_2)	Fuel combustion (especially high-sulfur coal); electric utilities and industrial processes; and natural sources such as volcanoes.	Aggravates asthma and increased respiratory symptoms. Contributes to particle formation, with associated health effects.

Source: "Our Nation's Air: Status and Trends through 2010," United States Environmental Protection Agency, EPA-454/R-12-001. February 2012. PDF page 3. Available at www.epa.gov/airtrends/2011/report /fullreport.

POLLUTION CONSEQUENCES DELAYED
IN SPACE AND TIME

As discussed above, ozone is not directly emitted by burning; instead, it is formed by the interaction between nitrogen oxides and volatile organic chemicals present in the atmosphere. Similarly, sulfur oxides and nitrogen oxides can interact with chemicals in the atmosphere to form acid compounds that contribute to acid rain. It is estimated that 62 percent of nitrogen oxides comes from transport, while 31 percent comes from electric power generation. In contrast, 67 percent of sulfur oxides comes from electric power generation, but that is predominantly the result of burning high-sulfur coal.

Acid Rain

During the 1960 and 1970s, when people in North America became increasingly conscious of how human activities can have an adverse effect on the environment, acid rain began to emerge as an issue that was discussed in the popular press. It also received attention after causing political strain between the United States and Canada because Canadians felt that they were victims of fossil fuel burning in the United States.

The EPA describes "acid rain" as a broad term referring to a mixture of wet and dry deposition from the atmosphere which contains higher-than-normal amounts of nitric and sulfuric acids. The precursors of acid rain originate in natural sources, such as volcanoes and decaying vegetation, as well as emission of sulfur dioxide (SO_2) and nitrogen oxides (NO_x) resulting from energy consumption. Acid rain occurs when these gases react in the atmosphere with water, oxygen, and other chemicals to form various acidic compounds, resulting in a mild solution of sulfuric acid and nitric acid. When sulfur dioxide and nitrogen oxides are released into the atmosphere, prevailing winds blow these compounds across state and national borders, sometimes over hundreds of miles. Acid rain largely results from the burning of fossil fuels with high-sulfur content.

Acid rain causes acidification of lakes and streams. It degrades sensitive forest soil and harms trees at high elevations. Acid rain also accelerates the decay of building materials and paints, including buildings, statues, and sculptures. Harvard University is said to cover some of its sculptures during the winter to avoid damage from acid snow.[49]

The effect of acid rain and snow can be quantified by measuring pH values in rivers, lakes, and streams. The "p" in pH is an abbreviation for the German word *potenz,* meaning power, and the "H" stands for hydrogen, so pH value for a particular solution describes the potential or power of hydrogen in that solution. In practice, pH is a logarithmic measure of hydrogen ion concentration, originally defined by Danish biochemist Søren Peter Lauritz Sørensen in 1909 using the following equation:

$$pH = -\log[H^+]$$

Pure water has a pH very close to 7 at 25°C. Solutions with a pH less than 7 are said to be acidic, and solutions with a pH greater than 7 are basic or alkaline. To appreciate the pH scale, it is useful to consider the pH of some common substances. Battery acid has a pH below 1. Lemon juice has a pH between 1 and 2, and vinegar has a pH between 2 and 4. Sea water has a pH of approximately 8. Baking soda has a pH of around 9, and milk of magnesia and the Great Salt Lake have a pH of approximately 10. Ammonia has a pH of 11, and liquid drain cleaner has a pH of almost 14. Most lakes and streams have a pH between 6 and 8, although some lakes are naturally acidic even without the effects of acid rain.

The impact of acid rain depends on the geology of the region onto which it falls. For example, alkaline soil neutralizes acid rain. Desert regions with little clay are often alkaline, which is why the Great Salt Lake is so alkaline. In such regions, acid rainfall may actually prove beneficial.

---

49. See en.wikipedia.org/wiki/Harvard_Bixi.

In contrast, acid rain can have a very negative effect in regions not characterized by alkaline soil. In the context of discussions of acid rain, "buffering capacity" refers to a soil's ability to neutralize acid rain. Thus, the negative effects of acid rain occur primarily in watersheds whose soils have a limited buffering capacity. In the United States, these regions include the Adirondacks and Catskill Mountains in New York State, the mid-Appalachian highlands along the East Coast, the upper Midwest, and mountainous areas of the western United States.

Many lakes and streams examined in the National Surface Water Survey suffer from chronic acidity. The survey investigated the effects of acidic deposition in over 1,000 lakes larger than ten acres and in thousands of miles of streams believed to be sensitive to acidification. Of the lakes and streams surveyed, acid rain caused acidity in 75 percent of the acidic lakes and about 50 percent of the acidic streams. Some lakes now have a pH value of less than 5. One of the most acidic lakes reported is Little Echo Pond in Franklin, New York, which has a pH of 4.2. Approximately 580 of the streams in the Mid-Atlantic Coastal Plain are acidic primarily because of acidic deposition. In the New Jersey Pine Barrens, for example, over 90 percent of the streams are acidic, which is the highest rate of acidic streams in the nation.[50] Emissions from U.S. sources also contribute to acidic deposition in eastern Canada. The Canadian government has estimated that 14,000 lakes in eastern Canada are acidic.[51]

The impact of nitrogen emission on surface waters is also critical. Nitrogen is an important factor in causing eutrophication (oxygen depletion) of water bodies. The symptoms of eutrophication include blooms of algae (both toxic and nontoxic), declines in the health of fish and shellfish, and loss of sea grass beds and coral reefs. According to the National Oceanic and Atmospheric Administration, these conditions are common in many coastal ecosystems. These ecological changes reduce the

50. See www.epa.gov/acidrain/effects/surface_water.html.
51. Ibid.

availability of seafood, create a risk of consuming contaminated fish or shellfish, and detract from the appeal of tourist destinations.[52]

Global Warming

Scientific evidence suggests that the burning of hydrocarbons has increased the level of carbon dioxide in the atmosphere, resulting in significant climate changes that are often colloquially termed *global warming*. An immense literature is available on this topic, but many basic scientific issues remain unclear, so I will not attempt to discuss the issue in this book.

RELATIONSHIP BETWEEN ENERGY USE AND POLLUTION

As Figure 10.1 shows, not all of the emissions of pollutants are associated with energy use. Figure 10.2 shows the fractional contributions to pollutants divided by source, where the sources are divided into three categories: (1) fuel combustion in power generation (red); (2) transportation (blue); and (3) and other non-energy-use-related activities (ivory). For pollutants where ivory dominates, such as PM10, PM2.5, and ammonia, moving to a 100 percent renewable future will not have a large effect on atmospheric pollution.

In contrast, sulfur dioxide is absolutely dominated by power plant emissions, and all of the rest are dominated by transport. Thus, with our present mix of primary sources for fossil fuel–burning electric power generation, completely switching to renewables would largely eliminate sulfur dioxide emission but would not have a much smaller impact on all other emissions.

As was shown in Figure 2.5, at present, coal provides approximately twice as much energy for electricity generation as does natural gas. The

52. See www.epa.gov/acidrain/effects/surface_water.htmlinations.

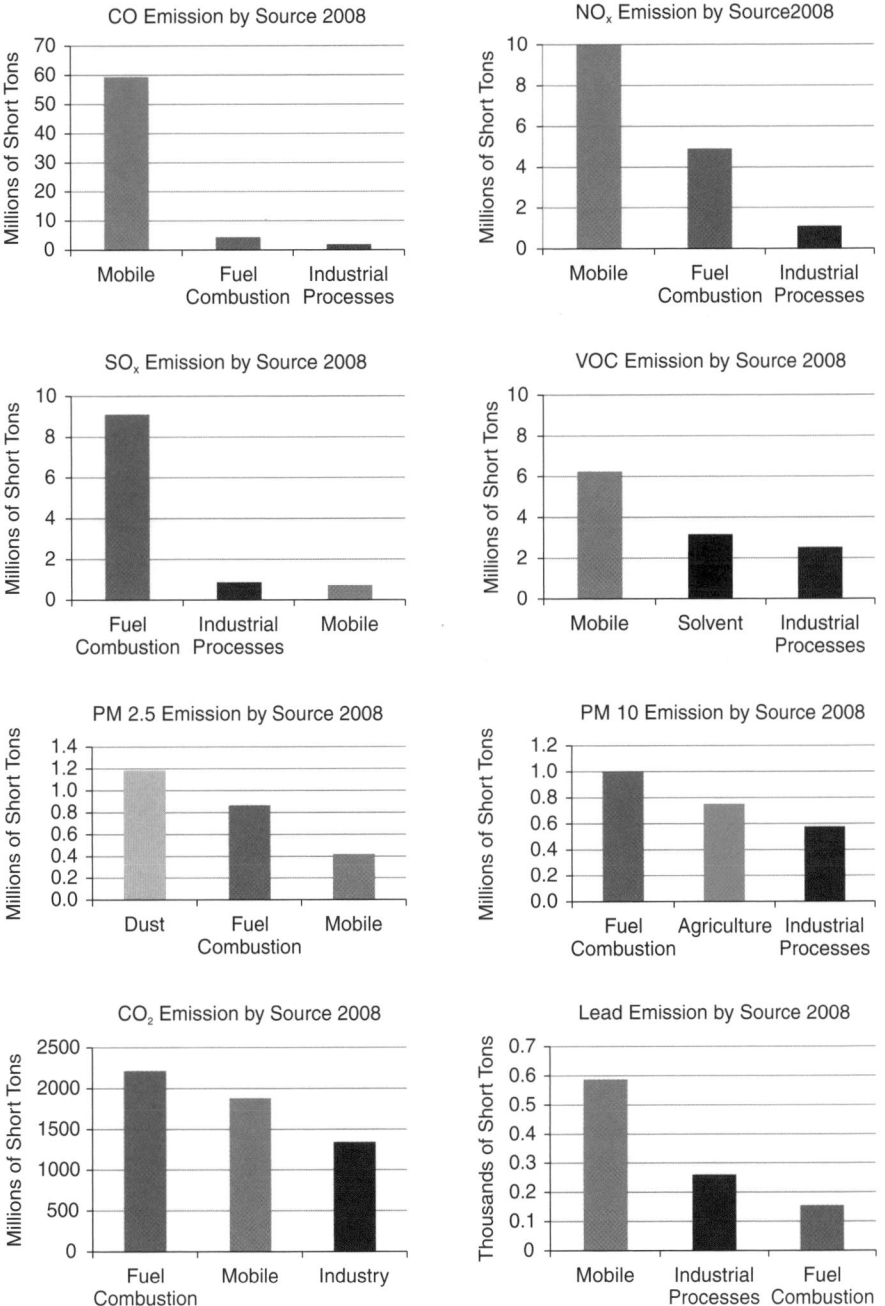

Figure 10.1 Short tons of emission from the top three sources of emission for eight pollutants based on EPA 2008 data. In every category except PM 2.5, emission is dominated either by electricity generation or transport (i.e., mobile). This information is based on tables that I generated from data at www.epa.gov/ttn/chief/net/2008inventory.html. More recent data is now available at www.epa.gov/ttn/chief/net/2011inventory.html. Information on emission trends is available at www.epa.gov/ttn/chief/trends/index.html.

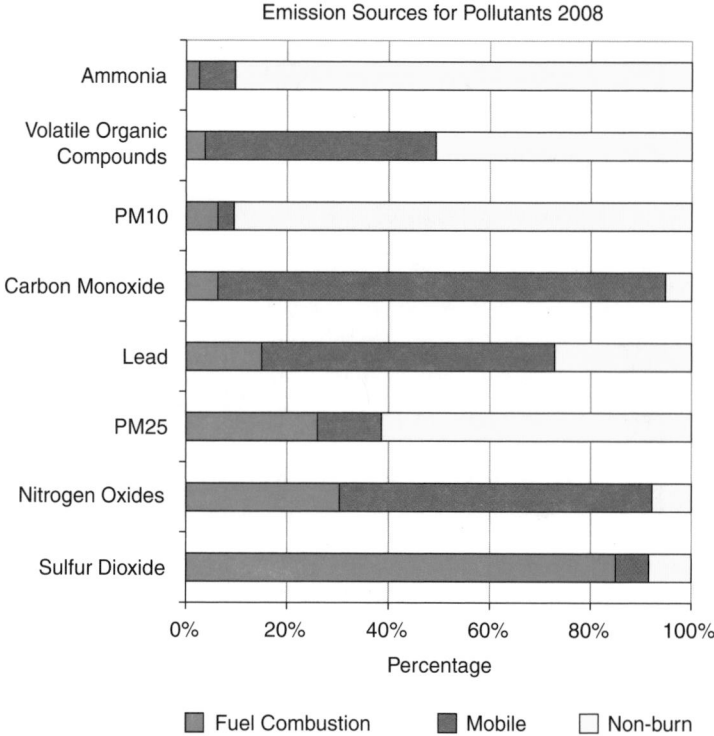

Figure 10.2 Fractional contributions of fuel combustion, transport, and nonburning interactions to various types of emission rank ordered by the contribution of fuel combustion, which is basically electric power generation. This information is based on tables that I generated from data at www.epa.gov /ttn/chief/net/2008inventory.html. More recent data is now available at www .epa.gov/ttn/chief/net/2011inventory.html. Information on emission trends is available at www.epa.gov/ttn/chief/trends/index.html.

very different pollution profiles for the two sources mean that power plant emissions from coal-burning plants produce much more than twice the emissions of the natural gas–burning plants. As shown in Figure 10.3, switching to 100 percent natural gas would also eliminate sulfur dioxide generation, without requiring any switch to renewables, because the sulfur dioxide coming from power plants is really all coming from coal-burning power plants. The story is the same with particulates and mer-

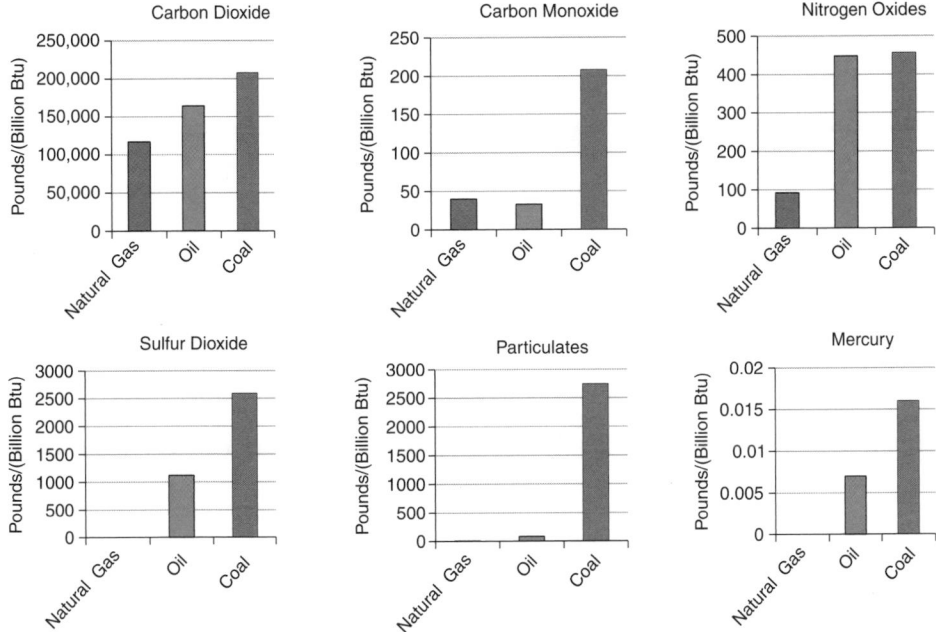

Figure 10.3 Emissions associated with fossil fuel burning expressed in pounds of pollutant per billion British thermal units of energy. In all categories, coal has the largest emission and natural gas has the smallest. Coal emits almost twice as much carbon dioxide as natural gas, four times more nitrous oxides and sulfur dioxides, and over two times more particles and mercury. U.S. Energy Information Administration Natural Gas Issues and Trends 1998. Information available at www.eia.gov/pub/oil_gas/natural_gas/analysis _publications/natural_gas_1998_issues_trends/pdf/it98.pdf, table 2, p. 58.

cury. Switching would also decrease nitrogen oxides by a factor of five. Drastic decreases in nitrogen oxides and sulfur oxides would greatly reduce acid rain, but in the United States nitrogen oxides emission is dominated by mobile sources, so eliminating coal-burning plants would not eliminate nitrogen oxides. Natural gas–burning plants still produce carbon monoxide, but as Figure 10.2 shows, carbon monoxide emission is also dominated by mobile sources. Thus, if we use the statistics shown in Figure 10.2 to argue in favor of switching electricity production to renewables on the grounds that emissions will be greatly reduced, we are vulnerable to the counterargument that similar reductions could be obtained simply by switching from coal to natural gas. Furthermore, this process is occurring naturally as a result of market forces. Thus, the emissions-based argument for renewable electricity generation rests almost entirely on carbon dioxide and its impact on global climate.

Coal

Coal is not one particular chemical entity. Coal includes various hydrocarbons combined with various other chemicals. The content of coal varies enormously from place to place, but it is always complex. When coal is burned, carbon dioxide, sulfur dioxide, nitrogen oxides, and mercury compounds are released. Nitrogen dioxide emissions can be created by the burning of nitrogen contained in the coal, and they can be produced by reactions between the oxygen and nitrogen in the air that occur because the air is hot enough to overcome the potential barriers that prevent them from recombining at room temperature. In the United States, coal-fired boilers are required to have control devices to reduce the amount of emissions that are released. Despite these controls, the average emission rates in the United States from coal-fired generation are 2,249 pounds per megawatt-hours of carbon dioxide, 13 pounds per megawatt-hours of sulfur dioxide, and 6 pounds per megawatt-hours of nitrogen oxide.[53]

Natural Gas

Natural gas is composed almost exclusively of methane. As shown in Figure 10.3, on average natural gas–burning electrical power plants produce half as much carbon dioxide, less than a third as much nitrogen oxides, and 1 percent as much sulfur oxides as does a coal-burning plant. The average emissions rates in the United States from natural gas-fired

53. See www.epa.gov/cleanenergy/energy-and-you/affect/air-emissions.html.

generation are 1,135 pounds per megawatt-hours of carbon dioxide, 0.1 pounds per megawatt-hours of sulfur dioxide, and 1.7 pounds per megawatt-hours of nitrogen oxides. Incomplete natural gas burning results in some emission of methane into the atmosphere. Emissions of sulfur dioxide and mercury compounds from burning natural gas are negligible.[54]

Internal Combustion Engine Combustion

Though natural gas is used in some vehicles, usually urban buses, as Figure Intro.6 showed, the vast majority of the energy used in transportation is provided by petroleum, and petroleum-based energy generation is almost exclusively confined to transportation.[55]

Below is an outline of the chemical transitions that occur during petroleum burning:

$$Fuel + Air \rightarrow Unburned\ hydrocarbons + Nitrogen\ oxides\ (NO_x)$$
$$+ Carbon\ monoxide\ (CO) + Carbon\ dioxide + water.$$

If the chemicals released in burning are released in the atmosphere, they can have significant local and global effects now and into the future.

Figure 10.4 shows the fractional contribution of different sources to the emission of pollutants monitored by the EPA. In this figure, the pollutants are ordered according to the contribution of transportation. The majority of the emissions of lead, nitrogen oxides, and carbon monoxide are contributed by transportation, which also contributes almost half of the volatile organic compounds. Converting from cars using internal

54. Ibid.
55. The use of gasoline in motor vehicles can result in pollution even when it is not burned at all. Evaporation of fuel also makes a contribution. Almost everyone is familiar with the smell of gasoline. We can smell it only because it is evaporating—that is, spontaneously converting from a liquid to a gas. A considerable amount of hydrocarbon pollution results from evaporative emissions that occur when gasoline leaks or spills or when gasoline gets hot and evaporates from the fuel tank or engine.

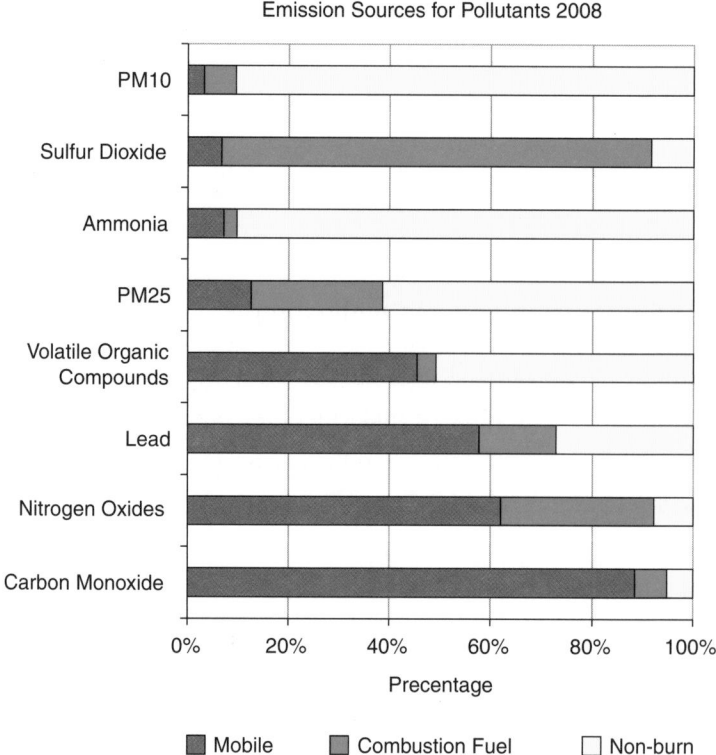

Figure 10.4 Pollutants rank ordered by the contribution of mobile sources, which is basically transportation. This information is based on tables that I generated from data at www.epa.gov/ttn/chief/net/2008inventory.html. More recent data is now available at www.epa.gov/ttn/chief/net/2011inventory.html. Information on emission trends is available at www.epa.gov/ttn/chief/trends/index.html.

combustion engines to cars using electric motors would eliminate emissions from cars, though of course the source of the electricity matters since fossil fuel–burning plants generate substantial emissions, as discussed earlier; however, if the electricity for the cars were provided by renewables, the impact on emissions would be enormously favorable.

It is important to note that automobiles are not the sole contributors to the mobile category. Despite automobiles outnumbering off-road vehicles,

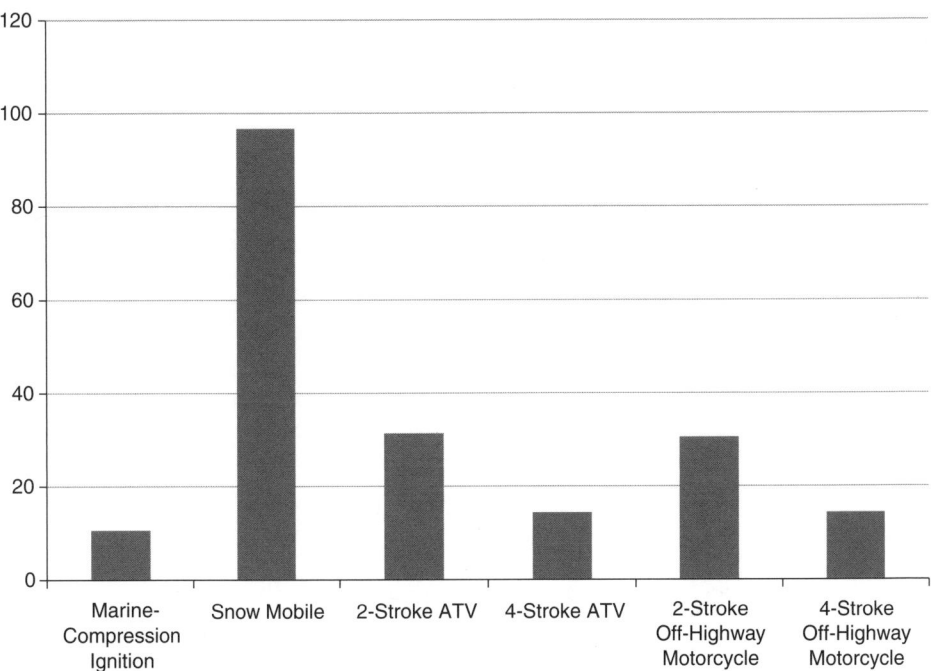

Figure 10.5 Emissions from various motorized vehicles graphed in terms of the number of cars that would produce approximately the same emissions as one single vehicle in one of these other categories. Dramatically, one snow mobile generates almost as much emission as a hundred cars. United States Environmental Protection Agency, Environmental Fact Sheet Frequently Asked Questions: Environmental Impacts of Recreational Vehicles and Other Nonroad Engines September 2001. Information available at www.epa.gov/nonroad /proposal/420f01030.pdf, document EPA420-F-01-030, p. 2.

some pollution categories are dominated by off-road vehicles because of their far less efficient two-stroke engines, which combine intake and exhaust in one cycle. Furthermore, the fuel often includes oil to provide engine lubrication. Two-stroke engines burn fuel less efficiently and emit more pollutants, as any user with a sense of smell can attest when the neighborhood landscapers are wielding their leaf blowers. Figure 10.5 shows the number of cars that would produce pollutants equivalent to one single recreational vehicle. Amazingly, the pollution from one snow-mobile is equivalent to the emissions of one hundred cars. Government regulations are encouraging the transition of off-road vehicles from two-stroke engines to four-stroke engines to reduce this problem.

Trends

Federal and state emission standards have had an enormous impact in reducing pollutants.[56] As a result of improvements in emissions standards, air quality in cities like Los Angeles has improved dramatically in the last fifty years, despite increases in the number of cars on the road. California's emissions standards are set by the California Air Resources Board, known locally by its acronym, CARB. Given that California's automotive market is one of the largest in the world, CARB wields enormous influence over the emissions requirements that major automakers must meet if they wish to sell into that market. In addition, several other U.S. states also choose to follow the CARB standards, which have also influenced E.U. emissions standards.[57]

As with many other energy issues, the most effective strategies for improvement may involve changes not in device technology but rather in how the devices are used. Thus, indirect emissions reduction strategies are also being tried. For example, carpooling is encouraged by creating "high-occupancy vehicle lanes," which are to be used only by vehicles

56. Detailed information of air quality trends in the United States is available at www.epa.gov/airtrends/. National and local information on each individual pollutant, along with comparison to national standards, is offered in graphical form and as spreadsheets. Of particular note is the graph at www.epa.gov/airtrends/images/y70_12 _lineStyles.png, which shows that from 1970 to 2008, while U.S. population, GDP, energy consumption, and miles driven has increased, overall emission of the six-tier major pollutants drastically decreased while the carbon dioxide emission increased almost proportional to energy use. Since 2008, carbon dioxide emissions have dropped significantly, while the emission of the other six major pollutants has continued to decrease.

57. See en.wikipedia.org/wiki/Emission_standard.

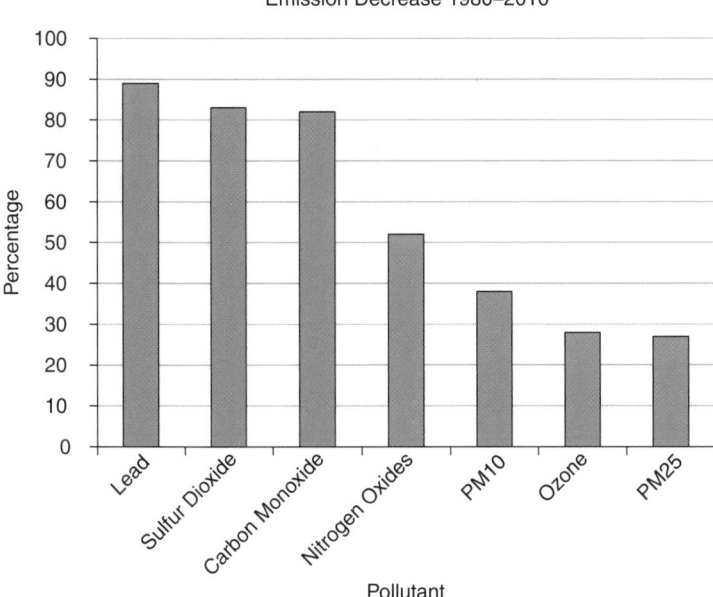

Figure 10.6 Emission decreases from 1980 to 2010. Information on emission trends is available at www.epa.gov/ttn/chief/trends/index.html.

carrying more than at least two passengers. Gas mileage standards also reduce vehicle emissions, as do municipal rules that limit idling. In 2003, the City Council in New York City passed a bill requiring the New York City Taxi and Limousine Commission to set aside a proportion of new taxi medallions to be granted to vehicles that use cleaner fuels. As an incentive for fleet owners to buy hybrids, in 2004 the Commission auctioned the first taxi medallions for hybrids at a discounted price of around $170,000 less than the regular medallion price of $400,000. In 2006, Mayor Michael R. Bloomberg announced that 254 of the 308 medallions to be auctioned by mid-2006 would be designated for hybrid and alternative-fuel cabs. By September 2012, hybrids represented 45 percent of the total taxi fleet in New York City.[58] They have proved to be popular with

58. See en.wikipedia.org/wiki/Hybrid_taxi.

owners and drivers because cabs spend much of their time either idling or in stop-and-go traffic. Hybrid cars use much less gasoline in such applications because hybrid braking allows the recycling of energy, and the "hotel power" required to run climate control and electronics usually does not require all of the power generated by an idling engine. Thus, hybrids provide higher profits to the cab operators, as well as improved air quality for city dwellers.

Even without making such a drastic change in technology, incremental engineering improvements and changes in fuel can make an enormous difference. Figure 10.6 shows the reduction in emissions between 1980 and 2010, ordered by the size of the decrease. Three of the pollutants have shown decreases in excess of 80 percent, and lead has shown a decrease of almost 90 percent. At present, only ozone and lead continue to exceed the national standards.

CONCLUSION

A Renewable Future

People are concerned with energy use for a variety of reasons, but most people are interested in the possibility of making improvements in one or more of the following: the price of delivered energy; the environmental consequences of energy use; and the social, political, and military costs associated with energy sales across national borders. Although Vice President Dick Cheney stated in 2001 that "Conservation may be a sign of personal virtue, but it is not a sufficient basis for a sound, comprehensive energy policy,"[1] technical improvements that maintain or increase the satisfaction of energy consumers while reducing negative consequences associated with energy use will be universally acclaimed as excellent contributions to a sound energy policy. Other changes in energy consumption require trade-offs. For example, measures that reduce undesirable emissions may increase the price of energy. Explosive debates can arise between people who value one benefit over another and feel that anyone who disagrees with them is selfish or foolish.

The present U.S. energy consumption system is riddled with inefficiencies, some of which offer opportunities to lower consumption without

1. See usatoday30.usatoday.com/news/washington/2001-05-01-cheney-usat.htm.

sacrifice. The inefficiencies fall into several major categories: (1) losses that can be improved by better engineering that reduces energy use that does not benefit people, (2) mismatches between energy supply and demand that either require maintaining rarely used excess capacity or operating generating systems below the capacity where their efficiency is optimal, and (3) fundamental scientific limits.

Losses that can be improved by engineering have been steadily declining as a result of both market forces and government regulation, which together have allowed the energy use per person to remain stable for around forty years while the energy per GDP dollar decreased by about a factor of two during the same period, as was shown in Figure 1.8. That figure also indicates that both trends are expected to continue for at least the next thirty years. Energy loss reductions in this category come in two types: passive and active. Passive improvements include better thermal insulation in buildings, more correct tire inflation, and improved aerodynamic design in transportation. Active improvements exploit advances in real-time computation to dynamically adjust energy consumption to meet the needs of the user with the smallest possible amount of energy. For example, in older aircraft, pilots directly controlled the parameters, such as fuel supply to the engines, that determine flight. In contrast, in newer aircraft, pilots do not directly control flight parameters, such as the amount of fuel flowing to the engines. Instead, pilots issue instructions, such as an increase or decrease to flight speed, which are interpreted by digital circuitry. The digital circuitry determines the control parameters, such as the fuel flowing to the engines, based not only on the pilots instructions but also on information provided by sensors located throughout the plane and on additional factors such as safety margins and fuel economy. Modern planes also have many more dynamically controllable parameters, such as variable inlet guide vanes and stator vanes, that allow planes to optimize performance by collectively adjusting all of the dynamical parameters to provide optimal performance. In contrast, older planes depended on static design parameters that may not be optimized for any set of flight conditions, but are chosen to provide acceptable performance over a range of conditions. Furthermore, when pilots controlled individual parameters, any optimization requiring adjustment of several parameters depended on the pilot making the appropriate calculation of how the overall performance would be affected by each individual parameter. As a result, systems with digitally controlled dynamic parameters provide better reliability, safety, and fuel economy than older systems.[2] These digital control systems also make piloting planes much easier.

At the more mundane level, several years ago I bought a new cell phone about which I was very excited; however, I was extremely disappointed to discover that the battery lasted only about twelve hours. Just as I was preparing to return to my old phone, an operating system update arrived and extended battery life beyond twenty-four hours. Obviously, the hardware in the phone was completely unchanged, and software adjustments alone halved the energy consumption by dynamically adjusting parameters without my noticing any loss in functionality. Thus, as both the plane and phone examples illustrate, vastly improved system performance and user satisfaction can be achieved by exploiting real-time sensing and rapid digital calculation to optimize operation by adjusting groups of performance parameters based on information provided by sensors.

Not only can smart devices optimize their energy use based on their own immediate individual circumstances; smart devices can be linked together, providing vast opportunities for energy reduction accompanied by lifestyle improvement. For example, apps can inform people about real-time locations of available parking places so that drivers don't have to circle endlessly seeking an empty spot. Similarly, drivers can get real-time

2. Nihad Daidzic, "FADEC advances allow better engine performance," *Professional Pilot,* March 2012. Available at www.propilotmag.com/archives/2012/March%2012/A2 _Digital_pl.html. Sanjay Garg "Aircraft Turbine Engine Control Research at NASA Glenn Research Center." Document number NASA/TM—2013-217821. Available at ntrs.nasa.gov/archive/nasa/casi.ntrs.nasa.gov/20130013439.pdf.

route information that optimizes for existing traffic conditions, decreasing bottlenecks where people are trapped in barely moving traffic. More futuristically, self-driving cars on highways could optimize their collective motion to minimize energy use for the whole group, just as schools of fish do.

The second category of energy inefficiency involves mismatches between demand and efficient supply that require either maintaining rarely used excess capacity or operating generating systems below the capacity where their efficiency is optimal. Such inefficiencies can also benefit from the linking of separate smart devices. Linking devices within an individual house can allow seamless adjustments to energy use within that house, such as switching between solar-heated water and water heated by electricity, providing hot showers of arbitrary length on demand while minimizing both the cost to the consumer and the power required from the electrical grid. In the absence of such smart control, either the user would have to limit showers to the available solar-heated water, or all of the water would have to be heated by the electrical grid. Smart grids can also optimize energy consumptions over many houses in a particular region. Smart meters allow electricity companies to monitor instantaneous electricity use for each costumer, enabling power companies to charge less for electricity consumed during times when demand is low. Such cost structures reward customers who voluntarily tailor their demand to match supply. Some consumers have embraced a more radical change: they are accepting meters that allow electricity companies to dynamically decrease the electricity drawn by the consumer during peak use times. Whether such forced consumption reductions at times of peak demand result in lifestyle sacrifices depends on how seamlessly smart devices can delay consumption in ways that are not apparent to the consumer. For example, if electric car battery charging requires one hour, most people would not mind if the power company controlled when that hour occurred as long as it occurred during the time when the user was asleep.

The third opportunity for reducing negative consequences associated with energy consumption requires a complete change in technology that replaces old devices suffering from fundamental scientific limits with new devices based on completely different scientific principles that have more favorable efficiency limits. For example, incandescent lightbulbs generate light only when they are so hot that almost all of the energy consumed by the bulbs goes into waste heat, resulting in enormous energy efficiencies. In contrast, LED lights do not generate light because they are hot. A 6-watt LED generates almost much light as a 60-watt incandescent lightbulb, but the LED generates less than one-tenth of the heat. Thus, replacing incandescents with LEDs removes a fundamental loss mechanism that could not be engineered out of incandescent lightbulbs. As a result, switching from incandescent lightbulbs to LED lightbulbs could reduce energy consumption by as much as 90 percent. Similarly, friction-based brakes dump energy into heat that cannot be recovered, whereas regenerative brakes stop cars by using the kinetic energy of the car to charge the battery. That energy stored in the battery can be used again to move the car forward. Thus, energy can be stored and recycled over and over again, allowing hybrid cars to get more than 50 miles per gallon in city driving, whereas even smaller nonhybrid cars that have been optimized for fuel economy get only around 30 miles per gallon in the city.

Most dramatically, the second law of thermodynamics requires that petroleum-burning vehicles and fossil fuel–burning power plants waste around 50 percent of the energy they consume. Of course, real systems are even less efficient. In particularly, gasoline-powered cars driven in urban areas have energy efficiencies of only around 15 percent, whereas electric motors can have efficiencies exceeding 90 percent, so replacing gasoline-powered cars with electrically powered cars would greatly reduce the amount of primary energy required to power transportation. In addition, electric-powered vehicles can exploit regenerative braking to recycle energy, providing additional efficiency.[3] The EPA already highlights the

3. In the end, it is not energy efficiency itself that is important in determining whether to make a radical technological change in primary energy sources; what is important is whether the change improves lives. Energy inefficiencies have very different implications for renewable and nonrenewable sources. Energy efficiency as it

overall energy efficiency of electric cars by assigning effective gas mileage ratings to them. These miles per gallon gasoline equivalent (MPGe) ratings specify the gas mileage required to equal the energy efficiency of that electric vehicle. For a Tesla Model S, which is a luxurious high-performance car, the values are around 88 MPGe in the city and 90 MPGe on the highway. The more modest Scion IQ EV offers an impressive 138 MPGe in the city and 105 MPGe on the highway.[4] The conversion has already begun and is likely to continue. The major impediment to widespread conversion is the limited range of electric vehicles; however, existing commercial electrical vehicles have ranges of up to 300 miles,[5] while U.S. vehicles travel on average less than 30 miles a day.[6] Rare long-range demand could be met by service stations that rapidly exchange depleted batteries for charged batteries or "zipcars" with longer ranges that could be used on demand. Thus, for most people, switching need not require lifestyle sacrifice and may even offer benefits, such as improved acceleration and quieter operation.

Simply converting from gasoline engines to electrical vehicles would eliminate oil imports, which would provide large economic and political benefits. The conversion could also greatly reduce the environmental damage caused by transportation if the electricity were generated by electrical power plants, which emit less than the gasoline engines they replace. If the electricity were generated by coal-fired plants, emissions

---

relates to wind and solar simply means that some wind and solar energy is left unharvested, which lowers possible environmental impacts associated with those sources. In contrast, fossil fuels whose energy is lost to waste heat must still be extracted, processed, and burned, with all of the negative environmental impacts associated with those activities.

4. Importantly, the losses in fossil fuel–burning systems implies that renewables need not replace the total energy in the fossil fuels; they need only to replace the fossil fuel energy that is not wasted.

5. Tesla motors offers a "supercharger" that in thirty minutes can provide charge for 170 miles of driving. See www.teslamotors.com/charging#/calculator.

6. In 2009, the average vehicle traveled 29 miles per day, which represented three trips with an average length of 10 miles. See nhts.ornl.gov/2009/pub/stt.pdf.

might actually increase. In contrast, generating the electricity using natural gas would clearly improve emissions; however, even gas-burning plants produce some pollutants. Most important, burning natural gas fundamentally requires the release of carbon dioxide,[7] which is believed to play a strong role in global climate change. Thus, the conversion from gasoline engines to electric cars is most favorable if the electricity is generated by systems that do not require combustion. Furthermore, all of these electrical cars would result in a vast increase in electricity demand, which is otherwise projected to be stagnant. The increased demand could be met by new renewable energy power plants that would not simply displace existing fossil fuel–burning plants that might otherwise continue to provide years of service. This is important because in the absence of increased demand, the increase in new renewable energy plants would be governed by the rate at which fossil fuel–burning plants are being retired. Otherwise, premature decommissioning of fossil fuel–burning plants would increase the effective cost of building a new renewable plant. As renewable energy becomes more developed, the replacement of retiring fossil fuel–burning plants with renewable energy plants may become routine, which eventually would result in a nearly 100 percent renewable energy economy that generates almost no emissions.

Overall, conversion to a 100 percent renewable energy economy would provide enormous benefits in both the developed and developing world. According to the United Nations Environmental Program, emissions from burning fossil fuels are associated with approximately one million premature deaths a year worldwide, while costing approximately 2 percent of GDP in developed countries and 5 percent in developing countries.[8] Even nonurban environments would benefit from a conversion to an energy economy based largely or entirely on renewables because consequences such as acid rain and climate change extend far beyond regions

---

7. This is discussed in detail in Appendix B.
8. See www.unep.org/urban_environment/issues/urban_air.asp.

where the original emissions were generated. Such improvements are so significant that if the costs of emissions were included, switching to renewables would be advantageous based on cost alone. In fact, including only a carbon cost is sufficient to make wind power the most inexpensive energy source, as was illustrated in Figure 5.2.

In this book, I have shown that it is quite technically and economically feasible for the United States to generate 100 percent of its average total energy consumption by using only either wind or solar power.[9] Allowing for a mix of wind, solar,[10] geothermal, and hydropower[11] makes the transition even easier.[12] Finally, efficiency increases resulting from the elimination of heat engines in power plants and cars make the transition easier still. The greatest challenge that wind and solar power face is that the sun and wind provide fluctuating energy generation that may not be correlated with fluctuating energy demands. Simply combining wind and solar power can reduce fluctuations since wind still blows in the dark, and solar power may be excellent on a windless day. Better still, seasonal variations in wind and solar can be smoothed out by mixing both and

9. Industrial processes, jet engines, and other applications requiring combustion could obtain their energy from burning biofuels.

10. Solar can include not only solar PV but solar thermal power. Solar thermal electricity generating systems have demonstrated energy storage where hot fluid can generate electric power through the night, but the most important solar thermal application may simply be heating water directly without generating electricity.

11. Hydropower could include not only traditional hydroelectric power but also power generated from waves or tides.

12. For some applications, like water heating or climate control, using the solar energy or geothermal energy to directly heat the water is both simpler and more efficient than generating electricity that is subsequently used to heat water. Furthermore, using solar or geothermal to provide a warmer heat reservoir for heat pumps would allow heat pumps to provide much more efficient heating when the outside temperature is very cold. Biofuel could provide enough energy for niche applications that require burning. At present, the United States produces around one million barrels of ethanol a day, and aviation uses around two million barrels of jet fuel per day. Thus, doubling our current ethanol production would provide sufficient jet fuel for aviation. See www.indexmundi.com/energy.aspx?country=us.

by pooling wind power from the entire United States. Furthermore, even though hydroelectric power can generate only a small percentage of total U.S. energy, hydroelectric dams store enough energy to meet 10 percent of the total energy 30 percent of the time or 30 percent of the energy 10 percent of the time.[13] Since hydroelectric power is available on demand, it could be used sparingly to buffer variations in the total production due to wind and solar that result in power generation that is insufficient to meet consumer demands.[14]

A smart energy system including diverse renewable sources that collect and disseminate energy over the entire continental United States might greatly regulate fluctuations in both supply and demand.[15] For example, electricity demand in New England peaks during the summer,[16] but electricity demand in Florida peaks during cold snaps in the winter.[17] In both cases, consumers are already requested to voluntarily curtail electricity use because of the strain on the electric power generating systems; therefore, these regions could benefit from pooling electrical resources from both areas. Similar improvements in the match between supply and demand can be achieved by pooling geographically distributed renewable sources, as a still-air day in Nebraska may not be a still-air day on Cape Cod. Additionally, sunrise in Cape Cod occurs almost four hours before sunrise in San Francisco. Thus, an energy grid extending over the entire United States would allow local fluctuations to be averaged out so that both the actual energy supply and demand approach their average

13. It is important to note that though hydropower as it exists now probably has enough energy to buffer fluctuations in renewable generation, it is not configured to deliver that energy rapidly enough to provide the required power buffer. The power generation capacity of existing hydropower plants would have to be significantly increased in order to provide the necessary power.

14. Geothermal power is another renewable source that can be obtained on demand.

15. The woodstove example in Chapter 1 illustrated that woodstoves are most energy efficient when they are functioning at their rated capacity. If less heat is required, less energy is consumed; however, the energy efficiency is also reduced.

16. See www.iso-ne.com/trans/celt/fsct_detail/.

17. See www.wjhg.com/news/headlines/Wet-Fl-238949671.html.

values despite instantaneous local differences between supply and demand.

Any remaining differences between supply and demand could easily be addressed if we had suitable large-scale energy storage, but at present we do not. Conversely, even with existing technology, the batteries for a hundred million electric cars would represent an enormous amount of tunable energy demand. If each car had a Tesla 85 kilowatt-hour battery that stores 300,000,000 joules, then all of the batteries would store around 10 percent of total U.S. daily energy consumption. In a more radical scenario, some of the energy stored in those batteries could be used to supply power when needed.

In sum, the total energy demand for the United States is predicted to be constant for approximately the next fifty years. Electricity generated by renewable energy can easily provide 100 percent of the average energy consumption of the United States during those next fifty years, virtually eliminating the negative environmental consequences associated with fossil fuel consumption.[18] Better still, since the disappearances of cyclic heat engines would increase energy efficiency, the actual total energy required from renewables would be less than half the current energy provided by fossil fuels. Finally, advances in computing allow energy supply and demand to be linked in real time, so they can be optimized to reduce energy consumption and cost without lifestyle sacrifice. In short, we could be on the cusp of an energy revolution, which might significantly improve the lives of almost everyone on earth if only we have the courage to seize the opportunity.

---

18. The few remaining niche applications that absolutely require combustion could be met by burning biofuels.

Appendixes

Acknowledgments

Index

A heat engine is a device that converts heat energy into mechanical work. Heat can be converted into work with 100 percent efficiency if the system is not required to cycle; however, in cycling systems, the conversion efficiency is limited by the first and second laws of thermodynamics. That efficiency limit is called the Carnot efficiency, in honor of Sadi Carnot, who first derived it. In a cycling engine, the engine returns again and again to some initial state at the completion of each cycle. If each cycle results in a net conversion of heat into work, then the same cycle can be repeated endlessly, converting heat to work forever.

Petroleum-powered cars and fossil fuel–burning electric power plants are examples of cycling heat engines. The first and second laws of thermodynamics place the same fundamental limits on all cyclic heat engines, no matter how the engine is configured, so I am free to consider a particular simple special heat engine and then generalize that result to the efficiency limits on internal combustion–powered cars and fossil fuel–burning power plants.

The amount of heat that is wasted depends on the maximum and minimum engine temperatures. In particular, I will show that if the peak engine temperature is T_{hot} degrees kelvin and the minimum temperature is T_{cold} degrees kelvin, than the efficiency of the cyclic heat engine must be less than

$$1 - (T_{cold}/T_{hot}).$$

Most engines cool using ambient air, which has a temperature of around 300 degrees kelvin; therefore, T_{cold} is approximately 300 degrees kelvin. Typical maximum engine and boiler temperatures are 600 to 700 degrees kelvin, so the maximum possible efficiency of petroleum-driven vehicles or fossil fuel–burning power plants is less than $1 - (300/700) = 57$ percent. Obviously, efficiency could be improved by increasing T_{hot} or decreasing T_{cold}. Considering a cold day in Siberia makes it clear why decreasing T_{cold} is difficult. A temperature of $-20°F$ is more than $50°F$ below the freezing point of water, but it is still 244 degrees kelvin. Thus, using T_{cold} of 244 degrees would increase the Carnot efficiency only to 64 percent, assuming that T_{hot} remained the same. In a system using water as a working fluid, T_{cold} cannot be lower than approximately 273 degrees kelvin, so a system using water has an efficiency limit of 61 percent. Similar issues arise with increasing T_{hot}. Aluminum melts at $660°C$, which gives an efficiency limit of 68 percent if T_{cold} is at room temperature.

Of course, real engines do significantly worse, but the important point is that no engineering improvement can ever offer a cycling heat engine that does better.[1] The energy loss in the heat engine is required in order to transfer heat energy to work. It is similar to a fee charged by a money changer: there is a loss involved, but no transfer could happen at all without the loss. Thus, if we want to transfer heat to work, we must pay the fee.

Qualitative Discussion of Cycling Heat Engines

In order to explain the Carnot limit, I will begin by working through a simple heat engine step by step. The basic principle of a heat engine is that the energy supplied by heat is converted into work. In the simple engine I am considering,

---

1. Interestingly, in 2012, natural gas–powered electrical plants had efficiencies exceeding 40 percent according to www.eia.gov/electricity/annual/html/epa_08_01.html. Some power plants use recycled waste heat from one engine to run a second lower-temperature heat engine. Running two heat engines is more expensive than running one, but the overall efficiencies of the combined system approach 60 percent. Like renewable energy systems, such combined cycle plants trade an initial capital investment against a long-term decrease in fuel costs, hoping that the fuel cost reduction will produce a useful return on the capital investment. "Waste heat" can also be recycled to provide climate control. For example, the streets and sidewalks of Holland, Michigan, are kept free of ice and snow by the waste heat generated by the electric power plant. Mary Schmich, "Is the future paved with heated concrete?" Chicago Tribune News, January 21, 2011, http://articles.chicagotribune.com/2011-01-21/news/ct-met-schmich -0121-20110121_1_sidewalks-downtown-oak-park-pat-zubak.

the heat interacts with gas molecules, resulting in an increase in the kinetic energy of the gas molecules. That increase in kinetic energy can then be converted into mechanical work. Thus, the gas acts as a broker that transfers heat energy to work. During the first half of the cycle, the gas acts as a perfect broker, transferring all of the heat energy to work. In contrast, during the second half of the cycle, the system must give up energy in order to return to the initial state. The gas still acts as a broker, but it is a broker that requires a fee.

Figure A.1 shows a simple heat engine consisting of a cylinder with a moveable piston at one end. Gas molecules are trapped in the cylinder between the left edge of the cylinder and the left face of the piston, where the piston is free to expand or contract. The total number of gas molecules always remains the same, but the volume occupied by the gas changes as the piston moves in and out. The system starts with hot gas at high pressure. The red rectangle indicates the cylindrical volume occupied by the hot gas, while the gray area indicates the remaining volume of the cylinder. The hot gas exerts a large pressure on the piston,[2] which causes the piston to move to the right, just as the steam in a tea kettle presses against the lid and pushes it up.

As the piston moves to the right, it can do work on an external object. For example, the pistons in a car engine exert force that makes the wheels on the car rotate. Of course, it requires energy to make the wheels turn. Similarly, it requires energy for the steam in a tea kettle to lift the lid of the kettle. In both cases, that energy is provided by the pressure of the hot gas. In the simple system I am considering, the gas in the piston is an ideal gas. If the gas in the cylinder is ideal, then there is a relationship between the gas pressure P, the gas temperature T in degrees kelvin given by

$$P = \alpha \ T/d,$$

where α is a constant and d is the distance between the left end of the cylinder and the left end of the piston.

If the piston moves to the right, the gas is doing work pushing against the piston. In contrast, pushing the piston to the left requires that something outside of the gas does work to compress the gas. The work done by the gas in the simple heat engine depends on the pressure difference between the gas in the cylinder

---

2. What really matters is not the gas pressure inside the cylinder but the difference between the gas pressure inside the cylinder and the gas pressure outside the cylinder.

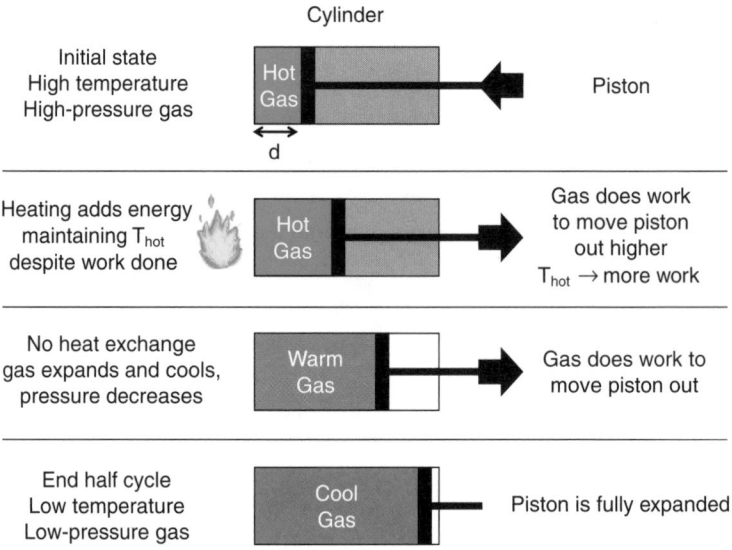

Figure A.1 Illustration of how heat energy can be converted into work by a piston during the first half of the heat engine cycle. The gas is always trapped between the left edge of the piston and the left end of the cylinder, but the volume occupied by the gas can change as the piston slides back and forth within the cylinder. The volume of the cylinder not occupied by the gas is shown in gray. The uppermost box illustrates the initial state in which a high-temperature gas is confined within a cylinder equipped with a piston that can slide back and forth. The next box down shows the cylinder expanding while it continues to receive heat. During this time the temperature of the gas remains constant, even though the gas is doing work, because the heat source keeps adding energy that exactly compensates for the work done by the gas. The third box down shows that the expansion of the piston continues even after the heat has been removed and the piston is thermally isolated. During this time, the internal heat energy stored in the gas provides additional work; however, since energy is conserved, that work done by the expanding piston comes at the expense of the kinetic energy of the gas. Thus, while the gas is thermally isolated, the gas cools as it expands. The cooling and expansion both reduce the gas pressure in the cylinder; therefore, at the end of the first half cycle, the gas in the cylinder is more expanded than it was in the initial state, and the gas temperature and pressure are lower than they were in the initial state. The flame image is from Mackie Drew.

and the gas outside of the cylinder, where larger differences correspond to more work.[3]

The simple cyclic heat engine under consideration here divides its operation into two halves. In the first half of the cycle, the expanding gas does work to push the piston out. For purposes of this discussion, I will connect the piston to the wheels on a car.[4] Thus, during the first half of the cycle, the pushing force from the gas will turn the wheels of the car and make the car move forward. In the second half of the cycle, something outside the cylinder must do work to compress the gas. In an automobile, the crankshaft compresses the gas in the cylinder by pushing the piston back in. Work is required to push the piston back in. That work consumes energy that otherwise could have been used to move the car forward. In order for a heat engine to power a car, the energy provided by the gas in the first half of the cycle must exceed the energy needed to push the piston back during the second half of the cycle.

Kinetic energy is the only energy in an ideal gas. Thus, for a thermally isolated ideal gas, conservation of energy implies that the kinetic energy of the gas must decrease if the gas is using energy to push the piston out and move the car forward. Similarly, if the car pushes back on the piston and compresses a thermally isolated gas, then conservation of energy implies that the kinetic energy of the gas must increase. This increase in kinetic energy can be seen when two people kick a soccer ball between them. If one person kicks the ball to another, and the second person simply stands still and lets the ball bounce off her stationary shins, then the ball does not move forward very fast; however, if she pulls her leg back and kicks in the direction opposite to the initial motion of the ball, then the forward motion of her leg will make the ball move away from her very fast. Similarly, if a gas molecule bounces off a piston moving toward the molecule, then the molecule will bounce off the piston with a larger velocity than the velocity with which it hit the piston, resulting in an increase in the kinetic energy of the gas molecule. Since for an ideal gas the temperature is proportional to the kinetic energy of the gas, compressing a thermally isolated gas increases its temperature, whereas the temperature of a thermally isolated gas decreases if the gas expands by doing work.

3. This is discussed in online Appendix 2 at http://thedata.harvard.edu/dvn/dv/HUP.

4. Real gasoline-powered cars use a more complicated system where the cylinder does not remain closed; however, one could make a car powered by a Carnot engine and that car would be correctly described by the discussion here.

Figure A.1 illustrates the first half of the cycle where the gas is doing work. That first half cycle can also be divided into two parts. In the first part, the gas is not thermally isolated: heat is applied to the gas to maintain the gas at a constant temperature T_{hot} while the gas expands by doing work to push the piston out. During first part of the half cycle, the heat provides just enough energy to keep the temperature of the gas constant even though the gas is doing work. In other words, on average, the gas molecules are simply transferring energy from the heat to the piston without an average net gain or loss. In the analogy between money and energy, the ideal gas is the perfect money changer: it is converting heat to work without taking any cut. Even though the energy of the gas is not changing, the pressure is, because $P = \alpha\, T_{hot}/d$, and d is increasing as the piston expands;[5] therefore, at the end of the first part of the first cycle, the temperature is still T_{hot}, but the pressure is lower than the initial pressure.

During the second part of the first half cycle, the heat is turned off and the cylinder is thermally isolated, so no heat can be added or subtracted. As long as the gas pressure is high enough, the gas will continue to expand and push the piston; however, since there is no heat to compensate for the work done by the gas, the gas will cool as it expands. Thus, during the second half of the cycle, the pressure decreases because d is increasing and because T is decreasing. As a result, at the end of the first half of the cycle, when the piston has completely extended such that it is pressed up against the far end of the cylinder, the gas temperature and pressure are both low; therefore, the first half of the cycle has transformed a hot, high-pressure gas occupying a small volume into a cold, low-pressure gas occupying a large volume, where that transition was accompanied by a delivery of work to the car attached to the piston. So far, the system can be 100 percent energy efficient. That sounds fabulous, but the problem is that the system cannot provide any more work to move the car forward because the piston is pressed against the far wall of the cylinder. Adding heat to increase the temperature and pressure of the gas will not move the piston, and

5. The pressure is going down as a result of the density of molecules in the gas going down, because the volume occupied by the gas is expanding while the number of gas molecules remains the same. As discussed in Chapter 4, on wind power, the pressure exerted by a gas is proportional to the number of gas molecules and the average velocity of the gas molecules. Thus, though the average velocity of the molecules in the gas remains the same because the added heat maintains a constant temperature, the pressure is going down because the density of the gas molecules is decreasing.

continuing to add heat can eventually make the cylinder explode. This is the problem with closed noncycling systems. They can be 100 percent efficient, but they cannot continue transferring heat energy to work forever. In order for the heat to do more work, the piston has to be pushed back. Pushing the piston back requires that some energy that could have been used to move the car forward must be redirected to move the piston back. This requirement that the cylinder return to its original state limits the energy efficiency of a heat engine, as I will explain below.

Compressing the gas requires work since the gas pressure will resist compression, just as pushing the lid on a tea kettle back down requires work if the steam is pushing it up. The higher the pressure in the cylinder, the larger the work required. If the car pushed back on the piston while it was not able to exchange heat with the outside world, then the gas inside the piston would heat up because each collision with the compressing piston increases the kinetic energy of the gas. Thus, the force pushing back on the car would increase as the gas compresses. If the gas were pushed back while the system was thermally isolated, then the pushing back would at best exactly undo the work done by the gas during the second part of the first half of the cycle. The former simply reverses the latter, and the temperature and pressure at the end of the first part of the second half of the cycle would be exactly the same as the temperature and pressure at the end of the first part of the first cycle. If the system remained thermally isolated while the piston was pushed fully back to its initial position, the gas in the cylinder would have a higher temperature and pressure than the initial gas. As a result, the car would have done more work to compress the piston than the expanding gas did to move the car. Such a system would not be a useful heat engine.

In contrast, if during the first part of the second half cycle the gas is cooled while the piston is pushed back, then the temperature of the gas can remain T_{cold}, even though the car is doing work to push on the gas, as illustrated in Figure A.2. During the second part of the second half cycle, the gas can be left thermally isolated, as illustrated in Figure A.2. While the gas is thermally isolated, compressing the cylinder increases the temperature and the pressure of the gas. As a result, at the end of the cycle, the gas is in exactly the same state that it was in at the beginning of the cycle—a high-temperature, high-pressure gas occupying a small volume in the cylinder. If the engine has done more work during the first half cycle than it required during the second half cycle, then the cyclic heat engine has successfully converted heat into work and this process can continue forever, as illustrated in Figure A.3.

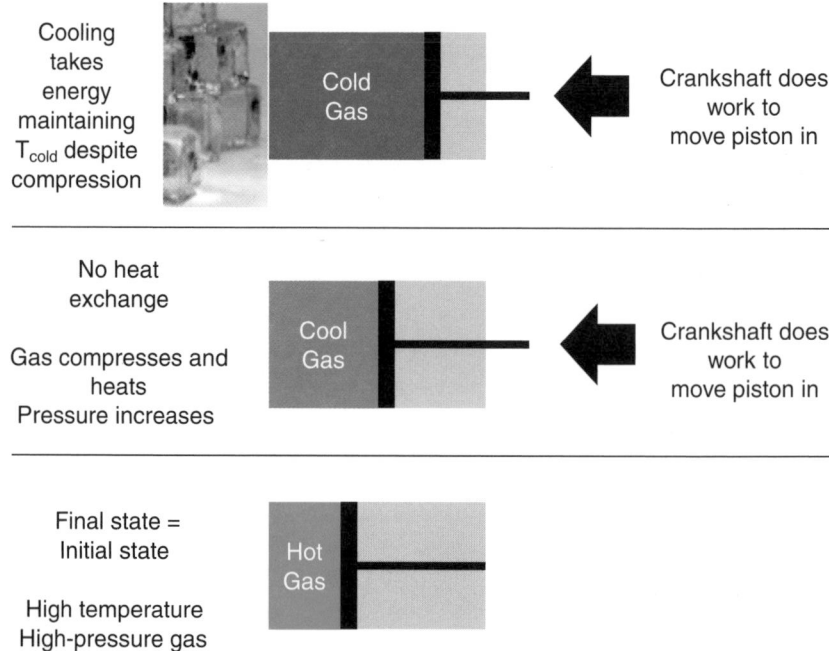

Crankshaft Does Work
Lower T_{cold} ® Less Work
T_{cold} = 0 kelvin ® 100% efficiency, but water freezes at 273 Kelvin

Cooling takes energy maintaining T_{cold} despite compression

Cold Gas

Crankshaft does work to move piston in

No heat exchange

Gas compresses and heats
Pressure increases

Cool Gas

Crankshaft does work to move piston in

Final state = Initial state

High temperature
High-pressure gas

Hot Gas

Figure A.2 Illustration of a cyclic heat can return to its initial state during the second half of the engine cycle, during which the engine dumps "waste" heat into the environment. The uppermost box follows the last box in Figure A.1, where the piston was thermally isolated and fully expanded. Once the piston is fully expanded, as shown in the last box in Figure A.1, the piston is connected to a cold reservoir that takes heat from the gas inside the piston while an external force does work to compress the piston. In a car, the external force could be obtained by taking energy away from the wheels. During this part of the cycle, the engine does work on the gas. If the gas were isolated, that work would increase the internal energy (temperature) of the gas; however, the cold reservoir is constantly removing heat. As a result, the temperature of the gas remains constant despite the work done by the engine. When the gas is partly compressed, the thermal contact between the piston and the cold reservoir is removed, as shown in the middle box. Even though the thermal connection is removed, the external force continues to do work to compress the gas. Again, this external force could be provided by taking energy from the car's wheels. Since the piston is thermally isolated, the work done by the external force is converted into the internal energy of the gas. Thus, the temperature and pressure of the gas increase as the external force continues to compress the gas. The lowermost box shows the system returned to the initial state shown in the uppermost box in Figure A.1. In this state, a hot, high-pressure gas occupies a small volume in the cylinder, and the part of the cycle shown in Figure A.1 begins again. The image of the ice cubes is from Darren Hester.

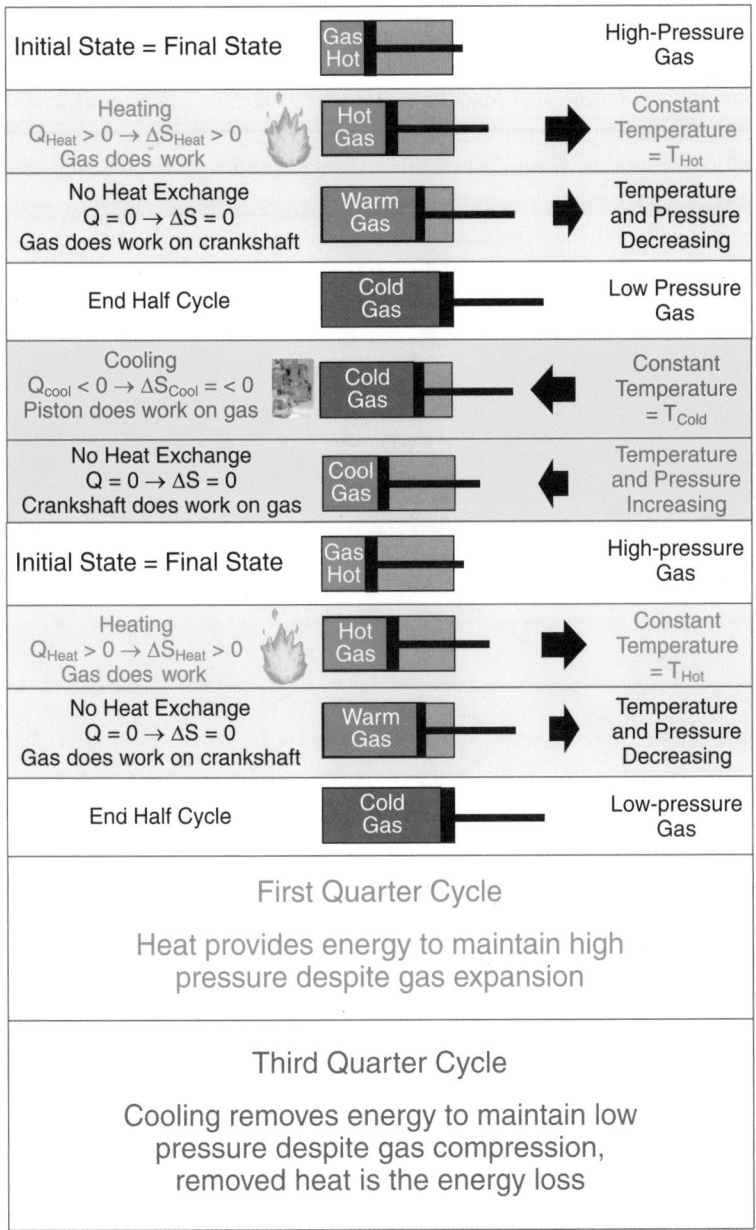

Initial State = Final State	Gas Hot	High-Pressure Gas
Heating $Q_{Heat} > 0 \rightarrow \Delta S_{Heat} > 0$ Gas does work	Hot Gas	Constant Temperature $= T_{Hot}$
No Heat Exchange $Q = 0 \rightarrow \Delta S = 0$ Gas does work on crankshaft	Warm Gas	Temperature and Pressure Decreasing
End Half Cycle	Cold Gas	Low Pressure Gas
Cooling $Q_{cool} < 0 \rightarrow \Delta S_{Cool} = < 0$ Piston does work on gas	Cold Gas	Constant Temperature $= T_{Cold}$
No Heat Exchange $Q = 0 \rightarrow \Delta S = 0$ Crankshaft does work on gas	Cool Gas	Temperature and Pressure Increasing
Initial State = Final State	Gas Hot	High-pressure Gas
Heating $Q_{Heat} > 0 \rightarrow \Delta S_{Heat} > 0$ Gas does work	Hot Gas	Constant Temperature $= T_{Hot}$
No Heat Exchange $Q = 0 \rightarrow \Delta S = 0$ Gas does work on crankshaft	Warm Gas	Temperature and Pressure Decreasing
End Half Cycle	Cold Gas	Low-pressure Gas

First Quarter Cycle

Heat provides energy to maintain high pressure despite gas expansion

Third Quarter Cycle

Cooling removes energy to maintain low pressure despite gas compression, removed heat is the energy loss

Figure A.3 Schematic of one and one half cyles of a heat engine. The first half of the cycle, where the expanding gas provides work, is highlighted in green. The second half of the cycle, where work must be done to compress the gas, is shown in orange. The image of the flame and the ice cubes were provided by Mackie Drew and Darren Hester, respectively. The system provides net work

(*Figure A.3 continued*)   because the work done when the hot gas pushes the piston out, allowing the car wheels to be pushed forward, is greater than the work done when the piston is compressed by the external force, which might be obtained by stealing energy from the moving car wheels. The work taken from the car is less than the work given by the car because the gas pressure is lower when energy is taken from the car than it is when energy is given to the car. This difference exists because the cold reservoir removes heat energy from the gas, allowing the pressure to be lower. That energy sent to the heat reservoir is "waste heat," which is lost to the system; however, that loss is required in order to get the piston back to its original position with a net energy contribution from the engine to the car to be greater than the contribution from the car to the engine. The lower the temperature of the cold reservoir, the less work the car has to do to compress the piston. Similarly, higher, hot-reservoir temperatures contribute more work to the car. These qualitative features are consistent with the result that the efficiency of an ideal engine is proportional to $(1 - T_{cold}/T_{hot})$.

As discussed above, the work done depends on the pressure of the gas: higher pressure requires more work. Since $P = \alpha\, T/d$ and T_{cold} is lower than any temperature achieved during the second half of the first cycle, at any given distance d, the gas pressure during the first part of the second cycle is lower than the pressure during the second part of the first cycle. Thus, the difference between the work done in the two parts is favorable. Furthermore, since the maximum pressure in the second half of the cycle is achieved when the gas is compressed back into its initial state, at every distance d, the pressure exerted by the gas during the second half of the cycle is lower than the pressure during the first half of the cycle. Thus, the work difference between the first part of the first half cycle and the second part of the second half cycle is also favorable. In sum, the requirements for the cyclic heat engine have been met: heat energy has been converted to mechanical work; however, not all of the heat energy was converted to mechanical work. The heat that was removed when the gas was cooled is lost forever. In the absence of cooling, the heat engine could not provide net work, so the heat loss required for the engine to provide net work limits the efficiency at which heat can be transformed into work: some of the heat must be lost to cooling, resulting in energy inefficiency.

Quantitative Discussion of Cycling Heat Engines

For the engine to operate in a cycle, the following conditions must be met: (1) the final temperature must be the same as the initial temperature; (2) the final piston position and gas volume must be the same as the initial piston position and gas volume; (3) the final entropy must be the same as the initial entropy; and (4) the final pressure must be the same as the initial pressure. Using just these logical requirements about cycling and applying conservation of energy and the thermodynamic definition of entropy, it is possible to determine the maximum possible efficiency of such an engine. The efficiencies of real engines must necessarily be worse. When the system is thermally isolated, the total energy of the system is conserved. In other words, the work done is exactly equal to the change in the kinetic energy of the gas. Thus, when considering the net energy change in the system, I don't need to consider the second part of the first half cycle or the second part of the second half cycle. The energy of the system changes only when heat is added or subtracted from the system; therefore, I need to consider only the first part of the first cycle, where the temperature is always T_{hot}, and the first part of the second cycle, where the temperature is always T_{cold}. If the system takes more energy from the heat than it gives back to the cold, then I have a useful heat engine that transforms heat into work.

The only time that energy is added to the system is when the gas is heated. Thus, conservation of energy, which is also the first law of thermodynamics, suggests that the work done in one cycle cannot exceed the energy put in, so

$$\text{Work} \le Q_{Heat}.$$

If one operated only the first half of the cycle, this would indeed be the work that one can extract; however, as discussed above, the second half of the cycle is required to return the system to its original state so that the the heat can provide more work. This means that the volume occupied by the gas must be decreased. In other words, all of the gas molecules that were distributed over the entire cylinder must be returned to the small volume they initially occupied. As discussed in online Appendix 2 (at http://thedata.harvard.edu/dvn/dv/HUP), this requires a reduction in the entropy of the system. It cannot happen spontaneously, as discussed in detail in online Appendix 2. The only way to reduce entropy is to actively remove heat from the system. This means that some of the energy expended to heat the gas must be wasted in order to return the gas to its initial state. This energy that must be removed is "waste heat."

Conservation of energy thus implies that the total work one can extract is

$$\text{Work} \le (Q_{Heat} + Q_{Cool}),$$

where $Q_{Cool} < 0$. That equation is great, but if we don't know anything about Q_{Heat} and Q_{Cool}, it is useless. Fortunately, we have the thermodynamic definition of entropy. Entropy is usually designated by the letter S. By definition, changes in entropy are related to changes in heat that occur at a temperature T by the equation $\Delta S = Q/T$. This equation can be rearranged to express the heat in terms of the change in entropy and the temperature, or $Q = \Delta S\, T$. Substitution into the equation above gives us the following:

$$\text{Work} \le (Q_{Heat} + Q_{Cool}) = \Delta S_{Heat}\, T_{Hot} + \Delta S_{Cool}\, T_{cold}.$$

We know T_{Hot} and T_{Cold} because we can measure them, but ΔS_{Heat} and ΔS_{Cool} are a mystery; however, we don't need to know them. All we need to know is that the total entropy change for a cycle is zero because in a cyclic system all of the variables at the beginning of the system must be the same as the variables at the end, and no entropy change occurs during the thermally isolated parts of the cycle since $\Delta S = Q$ and $\Delta T = 0$ because $Q = 0$. Thus, the change in entropy during heating must be equal and opposite to the change in entropy during cooling. In other words, $\Delta S_{Heat} = -\Delta S_{Cool}$. Substitution into the previous equation then gives us the following:

$$\text{Work} \le (T_{Hot} - T_{cold})\Delta S_{Heat}.$$

For now, what we are most interested in is not the total work but the efficiency of the system. We want to know how much work we can extract if we invest $Q_{Heat} = \Delta S_{Heat}\, T_{Hot}$ in the system.

Thus we are interested in the following:

$$\text{Work}/Q_{heat} \le [(T_{Hot} - T_{cold})\Delta S_{Heat}] / [(\Delta S_{Heat}\, T_{hot})] = 1 - T_{cold} / T_{Hot}.$$

This is Carnot's efficiency theorem.

Electricity from Heat

In the United States, the contributions of all renewable energy sources are dwarfed by electrical energy that is generated by heating a fluid that does work to rotate the shaft in an electrical generator. Thus, the vast majority of modern power plants provide power using the basic principles that make the lid of a teapot lift up when the water inside the teapot boils. This effect is discussed in detail in online Appendixes 2 and 3 at http://thedata.harvard.edu/dvn/dv/HUP. In this appendix, we will consider only the general conversion of heat to electricity. The required heat can be provided by sunlight, nuclear reactions, or combustion; however, in the United States the heat is dominantly provided by combustion.

Although people intuitively understand that burning produces heat, the idea that heat is a form of energy is relatively new. It developed in the mid-1800s along with the general idea that energy is conserved. Two of the major contributors to this development were James Prescott Joule and Julius Robert von Mayer. Both are known as physicists, but they also practiced other professions. Von Mayer was a doctor[1] and Joule was a brewer.[2] There is some controversy about the contributions made by the two men.

Von Mayer developed many of his ideas during three years he spent as a ship's physician in the Dutch East Indies. He noted that in the tropics the blood

---

1. See www.whonamedit.com/doctor.cfm/3561.html.
2. See www.aps.org/publications/apsnews/200912/physicshistory.cfm.

in veins is redder than it is in colder climates.[3] He realized that oxidation, the transfer of electron bonding between atoms that results in decreases in energy, is the primary energy source for living organisms. He is most famous for developing the correct quantitative relationship between work and heat, though he is also given substantial credit for proposing that energy is a conserved quantity.[4]

In contrast, Joule's studies were motivated by the possibility that steam engines could be replaced by battery-driven motors, but he discovered the problem that continues to plague battery-powered vehicles today—that the batteries required to run the motor for a given time weigh more and take up much more space than the coal required to run the motor for the same time. Of course, cars now use gasoline rather than coal, but the basic issue is the same.[5]

Joule conducted rigorous experiments. He was so proud of the results of those experiments that his gravestone is inscribed with the number 772.55, which was his most accurate measurement of the mechanical equivalent of heat.[6] In the course of his work, he noted that running electrical current through metal heats the metal, which we now know is the result of friction that electrons experience as they move through the metal. He also did experiments that showed that mechanical work can heat water, and he measured the amount of mechanical work required to compress gas in a piston. Together, all of his work suggested that energy is conserved and that heat is a form of energy. In the end, Joule's contributions were considered so important that the Joule is the standard quantity of energy in the meters-kilometer-second system of measurement in honor of Joule's achievements.[7]

The full understanding of the relationship between heat and energy required an appreciation for the connection between the temperature of an object and the average energy of the molecules making up the object. In 1738, Daniel Bernoulli correctly proposed that gases consist of individual moving particles and that the pressure that gases exert on surfaces is a result of collisions between the molecules and surfaces. Thus, a hot gas is hot because the molecules in it

3. See www.uh.edu/engines/epi722.htm.

4. See www.chemgapedia.de/vsengine/glossary/en/vonmayer_00045julius_00045robert_000451814.bio.html.

5. See www.aps.org/publications/apsnews/200912/physicshistory.cfm, paragraph 3.

6. See www.findagrave.com/cgi-bin/fg.cgi?page=gr&GRid=19866591. The website provides images of Joule and his gravestone.

7. See www.aps.org/publications/apsnews/200912/physicshistory.cfm.

are moving faster than the molecules in a cold gas. In other words, hot gases have more kinetic energy, as discussed in online Appendix 3. In particular, if T is the temperature in degrees kelvin and k is Boltzmann's constant, then $1/2$ m v^2, the kinetic energy of a gas molecule, is equal to $3/2$ kT.[8]

The relationship between heat and kinetic energy was developed by Rudolf Julius Emanuel Clausius, who was a German physicist and mathematician. In the period between 1850 and 1865, Clausius developed the relationship between heat and kinetic energy to explain how mechanical energy can be converted into heat. His work established the principle of energy conservation that had been correctly proposed by von Mayer and Joule.[9] Conservation of energy became established as the first law of thermodynamics, which is often used to place constraints on the maximum possible energy that can be extracted from different primary fuel sources.

The second law of thermodynamics also plays an important role in discussions of energy efficiency. The second law of thermodynamics states that the entropy of a system must either remain the same or increase unless heat is extracted from the system.[10] The second law places strong constraints on energy extraction systems that work in cycles, such as the gasoline engines in cars and the steam turbines in electric power plants, as discussed in Appendix A.

The illustration in Figure B.1 compares a hydroelectric power system with a heat-based power system.

In hydroelectric systems, the kinetic energy required to turn the shaft is generated from the gravitational potential energy that exerts a pressure on the water at the bottom of the dam that forces the water out through a nozzle at a high speed. The high speed of the water flowing through the nozzle is associated with a large kinetic energy. In steam generators, the large kinetic energy in the water flowing through the nozzle is created by the heat that induces water to expand and exert a pressure on turbine blades. It is the same principle that makes a tea kettle whistle as the heated steam rapidly moves through the hole

8. See /galileo.phys.virginia.edu/classes/252/kinetic_theory.html. Michael Fowler is a Professor at the University of Virginia who provides a wealth of beautiful, clear information about physics related topics.

9. See www-history.mcs.st-and.ac.uk/Biographies/Clausius.html.

10. Entropy is discussed in more detail in online Appendix 3, section 3.5, at http://thedata.harvard.edu/dvn/dv/HUP.

Hydroelectric Power

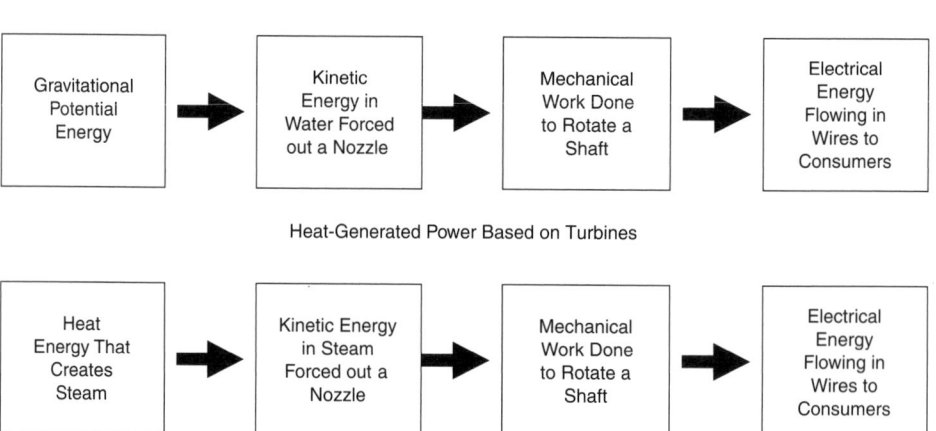

Figure B.1 Schematic comparison of the steps in hydroelectric power generation and the steps in heat-based power generation.

in the lid of the kettle. Furthermore, like hydroelectric systems, steam turbines exist in two modes—impulse generators and reaction generators.

The modern steam turbine was invented in 1884 by Sir Charles Parsons. His first model generated approximately 7.5 kilowatts of electricity,[11] which would be enough to power approximately seven modern U.S. households. A schematic of his system is shown in Figure B.2. The fluid flows in the clockwise direction, as indicated by the white arrows. The inside of the building is indicated in gray and the outside in green. To understand how a cycle works, consider the water after the condenser as the starting point. This water is cool and its pressure is low. It is shown in blue to indicate that its temperature is low. It moves through the pipes to the pump. The pump uses electrical energy to increase the pressure on the water. The energy required to operate the pump is supplied from outside the system, as indicated by the small pink arrow. As discussed in Appendix A, compressing the gas also increases the temperature, so the water color is shown as purple. The high-pressure water flows to the boiler, where it is heated. Thus, the environment gives heat to the fluid, as indicated by the large orange arrow at the top of the figure and pointing toward the boiler. The steam is

11. See www-g.eng.cam.ac.uk/125/1875-1900/parsons.html.

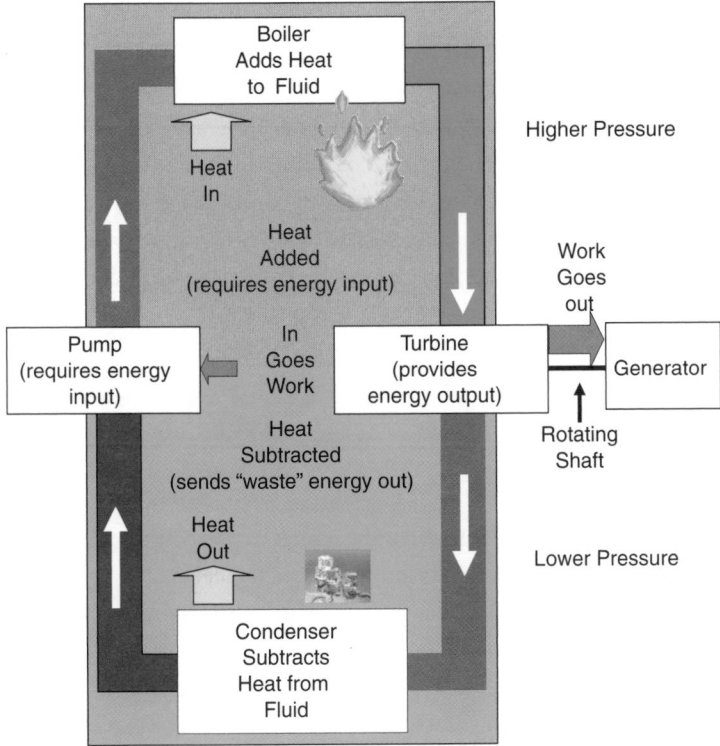

Figure B.2 Schematic of a heat-based turbine system used to generate electricity. The heat that is added in the boiler increases the pressure of the vapor. The high pressure of the vapor escaping from a nozzle imparts kinetic energy to the turbine blades, which turn the shaft in the generator. After the vapor leaves the turbine, it is at a lower pressure and temperature. The vapor passes through the condenser. The environment draws heat from the condenser, where the vapor is converted back into a liquid. The liquid leaves the condenser at a lower temperature and pressure. The liquid then passes through a pump, which does work to increase the pressure and temperature of the gas just before the gas enters the boiler and is heated once again. Image of the flame by Mackie Drew.

shown in red to indicate that it is very hot. Heating the water creates steam whose molecules are moving at high velocities. The hot steam moves through the pipe shown in red until it is pushed out of nozzles in the turbine. The inter- action at the turbine converts some of the high-kinetic energy of the water to the turbine blades, which are attached to a generator shaft. Thus, the energy from the hot fluid is converted to mechanical work that leaves the system, as indicated by the large magenta arrow pointing toward the generator. The lower energy steam has much lower pressure and a lower temperature, so its color is shown in purple. It leaves the turbine and flows to the condenser. In the condenser, heat flows from the steam to the outside, as indicated by the small orange arrow pointing away from the condenser. The flow of heat from the steam cools the steam and converts it back into water. The fluid leaves the condenser as cool low-pressure water, and then the cycle starts again. In an ideal system, then, the water completes the cycle in exactly the same state that it started and is simply an intermediary that converts the heat into electrical energy.

Thus, the system converts heat energy into mechanical work, which is converted into electrical work by the generator. In each cycle, the fluid takes more heat from the environment than it gives back, which is why the orange arrow pointing in to the boiler is larger than the orange arrow pointing out of the condenser. Similarly, the system gives out more mechanical energy than it takes in, which is why the magenta arrow indicating the energy from the outside that is required to do mechanical work on the fluid is smaller than the magenta arrow from the turbine to the generator, which indicates the mechanical work done by the fluid. In an ideal system, this difference in heat energy is balanced by the difference in mechanical energy. The energy efficiency limits in such systems are discussed in Appendix A.

Geothermal Heat to Electrical

It is possible to provide the heat required to run a boiler-based electrical power plant using heat sources other than combustion. For example, hot springs can also heat the gas in a gas turbine generator. Geothermal systems are considered to be renewable energy sources since radioactivity within the earth will continue to generate heat for millions of years. An important issue in geothermal heating is the required depth of the well. As anyone who has dug a garden in New England knows, digging a few feet into the earth does not result in a significant increase in temperature; however, in places with large volcanic activity, molten lava is available at the surface of the earth without digging at all. Furthermore, in many regions, natural hot springs bring water to the surface without digging. Finally, if one digs deeply enough straight toward the center of the earth from the United States, one may get not to China but to a high-temperature region inside the earth. Figure B.3 shows the available temperature for a well that is

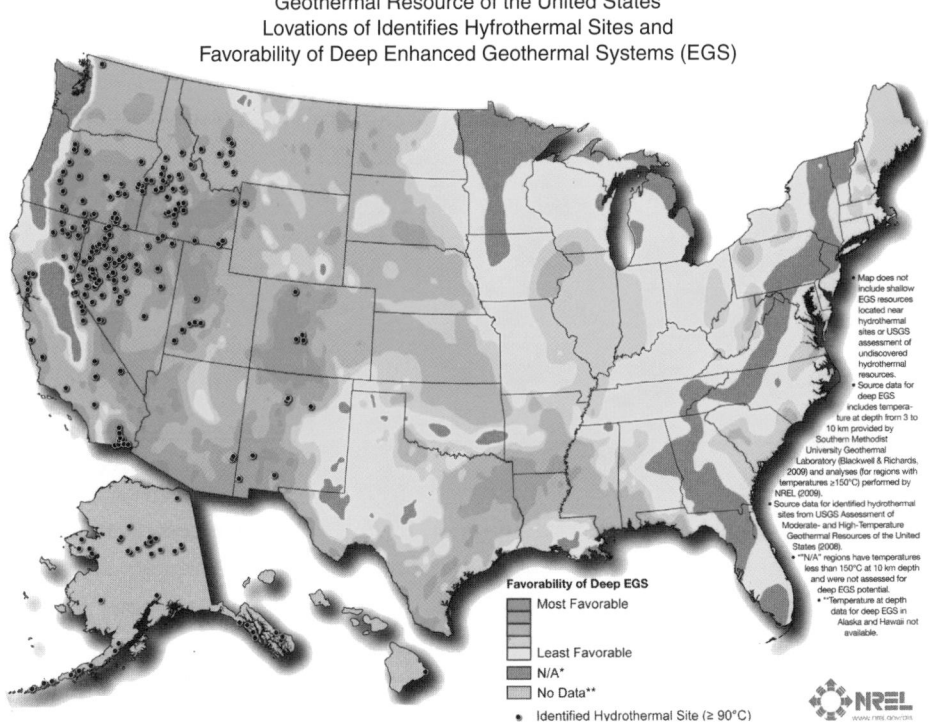

Geothermal Resource of the United States
Lovations of Identifies Hyfrothermal Sites and
Favorability of Deep Enhanced Geothermal Systems (EGS)

Figure B.3 National Renewable Energy Laboratory map of the availability of deep geothermal energy in the United States. Red is most favorable, and yellow is least favorable.

drilled approximately 4 kilometers into the earth, that is, more than 2 miles deep. The map shows that like wind and solar power, the siting of geothermal electrical generating plants can make a large difference.

Below I consider geothermal-based electricity generation, but it is probable that except in particular select locations, geothermal heat will play a much larger role in climate control than in direct generation of electricity.

Geothermal Heat to Electricity Systems

Hot springs have been enjoyed for thousands of years. Hot springs are hot because they flow from regions inside the earth that are warmer than the surface. Paleolithic people were able to enjoy a nice hot soak long before the invention of indoor plumbing. By Roman times, hot springs represented a substantial business opportunity supported by vast engineering works. The English word *plumbing* is derived from the Latin *plumbum,* meaning lead. Thus, plumbing has its origin in the lead pipes used by Roman engineers to transport water. Hot springs are widely believed to have therapeutic effects, and they continue to generate a significant tourist trade. The entire city of Bath in the United Kingdom was designated a world heritage site in 1987 because of the baths that date from Roman times.[12] Bath still receives water at a temperature of 46°C (115°F).[13]

The approximately 50°C water that is readily accessible for geothermal energy is well suited to climate control; however, water at this temperature is less well suited to electrical power generation because the Carnot efficiency decreases with the temperature of the heat source. Present efficiencies, which vary between 5 and 15 percent, are comparable to present solar photovoltaic efficiencies.

Of course, energy efficiency is less important in geothermal generating plants than in fossil fuel–burning plants because the adverse effects of mining and burning are not an issue; however, the capital cost of power plants and the power that can be derived for each dollar of capital investment are important. Higher-temperature water can be obtained by drilling more deeply into the earth, but substantial costs are associated with drilling and maintaining such wells.

Though geothermal power emits less carbon and sulfur than coal burning, for example, some emissions are still associated with it. The emission rate depends strongly on the source; nevertheless, geothermal electrical power generation is developing rapidly. The Union of Concerned Scientists[14] and International Geothermal Association provide a wealth of information about the science underlying geothermal power.[15] The International Geothermal Association also offers information on past and current installed capacity.[16] In 2005, there were 9064 megawatts of installed power capacity in twenty-five countries.[17] In

12. See whc.unesco.org/en/list/428.

13. R. W. Gallois, "The formation of the Hot Springs at Bath Spa, UK." 2007. *Geological Magazine,* vol. 144, 741–747. Information available at www.dorsetgeologist sassociation.com/RWG/hotsprings/2007_Hot_springs.pdf.

14. See www.ucsusa.org/clean_energy/our-energy-choices/renewable-energy/how -geothermal-energy-works.html.

15. See www.geothermal-energy.org/geothermal_energy/what_is_geothermal _energy.html.

16. See www.geothermal-energy.org/geothermal_energy/electricity_generation.html.

17. See www.geothermal-energy.org/geothermal_energy/electricity_generation.html.

Iceland, geothermal power provides 30% of electricity generation, and in the Philippines and El Salvador it provides more than 25%.[18]

The IGA reports that as of 2013, 11.772 megawatts are online, generating 67,246 gigwatt-hours. This represents a 10 percent increase in geothermal power online between 2010 and 2013. The United States is the world leader in generation with 3389 megawatts of capacity,[19] but the total power generated in the United States is only slightly higher than the 2,080 megawatt nameplate power of the Hoover Dam,[20] and geothermal represents only around 0.3 percent of U.S. power generation; however, this represents approximately 30 percent of the total geothermal power generated in the world.

The World Bank's estimates of the cost of geothermal electrical power generation ranges from approximately $0.03 to $0.15 per kilowatt-hour, depending on assumptions, with cost decreasing significantly with increasing power plant size.[21] If these estimates are correct, then in some selected locations, geothermal systems could provide electricity at a cost less than fossil fuel burning; however, the difficulties associated with it are sufficient to make widespread general adoption unlikely. Both wind and solar power are being much more aggressively pursued as alternative sources of electrical energy.[22]

18. See en.wikipedia.org/wiki/Geothermal_electricity, accessed June 24 2014.

19. See en.wikipedia.org/wiki/Geothermal_electricity, accessed June 24 2014.

20. See www.usbr.gov/lc/hooverdam/faqs/powerfaq.html.

21. "Technical and Economic Assessment of Off-Grid, Mini-Grid and Grid Electrification Technologies Annexes," The World Bank Group Energy Unit, Energy, Transport and Water Department. September 2006. Tables A.25 to A28 on PDF pages A-26 to A27. Available at siteresources.worldbank.org/EXTENERGY /Resources/336805-1157034157861/ElectrificationAssessmentRptAnnexesFINAL 17May07.pdf. And a more recent source "Geothermal Handbook: Planning and Financing Power Generation," Energy Sector Management and Assistance Program, Technical Report 002/12. International Bank for Reconstruction and Development, World Bank Group. Table 1.7 on PDF page 42. Available at www.esmap.org/sites /esmap.org/files/DocumentLibrary/FINAL_Geothermal%20Handbook_TR002-12 _Reduced.pdf.

22. Geothermal may play a large role in directly providing heat, as it does in Iceland where "About 87% of the population enjoy central heating by geothermal energy at a price that is generally less than half of the comparable cost of oil or electric heating," according to www.icetradedirectory.com/english/industry_sectors_in_iceland/energy _in_iceland/.

Sunlight to Heat to Electrical

Solar photovoltaic systems, in which photon energy is converted directly into electrical energy, were discussed in detail in Chapter 5. At the beginning of this appendix, I considered how heat can be used to drive a turbine that generates electricity. In this section, I will consider only electric power generation in which heat from the sun can be used to drive a turbine. This was probably the sort of system that Thomas Edison envisioned in 1931 when not long before he died, the inventor told his friends Henry Ford and Harvey Firestone: "I'd put my money on the sun and solar energy. What a source of power! I hope we don't have to wait until oil and coal run out before we tackle that."[23]

It has long been known that sunshine warms objects. Von Mayer developed his theories of the relationship between heat and light partly based on observations of thermal effects in the tropics. Simple solar water heaters are used by campers and people with outdoor showers, but the high temperatures that are so desirable to achieve high Carnot efficiencies cannot readily be obtained unless the sunlight is focused. The idea of focusing sunlight to heat materials to high temperatures is very old. Legend has it that Archimedes used a "burning glass" to concentrate sunlight on the invading Roman fleet and repel them from Syracuse. The 1941 movie *Ball of Fire,* starring Barbara Stanwyck and Gary Cooper, featured a group of professors who took advantage of the idea to help free themselves from a group of gangsters. Less charmingly, unkind children were often caricatured using magnifying glasses to focus solar power onto unfortunate ants.

In solar thermal electrical generation systems, mirrors focus sunlight that is delivered to a large area onto a central point. This concentration of solar power makes it possible to create the steam required for the steam turbine systems described earlier. According to a July 29, 2009, article in MIT's *Technology Review,* "Though solar thermal generation systems have been pursued since the 1970s, a breakdown in government funding and incentives caused them to stall before they reached a scale of production large enough to drive down costs and allow

---

23. Reconsideration, *New York Times Magazine,* "Current Thinking" June 3, 2007 by Heather Rogers. Available at www.nytimes.com/2007/06/03/magazine/03wwln -essay-t.html?_r=0. The original source was not cited, but according to http:// message.snopes.com/showthread.php?t=31707, the source may be *Uncommon Friends: Life with Thomas Edison, Henry Ford, Harvey Firestone, Alexis Carrel, and Charles Lindbergh James Newton,* first Harvest/HBJ ed. (New York: Mariner Books), June 23, 1989, p. 31.

them to compete with conventional sources of electricity. 'It was a classic problem with solar. The market support to bring solar to high volume wasn't there,' according Ian Simington," chief executive of the solar division of NTR, a renewable energy company based in Ireland.[24]

Solar Towers

In one type of system, the energy from the sun is focused onto a tower by tracking mirrors. In 2008, BrightSource Energy dedicated its Solar Energy Development Center in Israel's Negev Desert. The installation features more than 1,600 heliostats that track the sun and reflect light onto a 60-meter-high tower. The concentrated energy is then used to heat a boiler atop the tower to 550°C, generating steam that is piped into a turbine, where electricity can be produced.[25]

Solar Trough Systems

An alternative to the tower system is the parabolic trough system. Parabolic mirrors have the wonderful property of focusing all light that falls on them into a single line. In a solar trough system, that single line at the focal point is occupied by a pipe that contains the working fluid for the system so all of the energy from the light that hits the trough is directed to that working fluid with no energy loss. This represents a compression of solar energy, but the total sunlight collected is still less than or equal to the sunlight falling on the collectors. The area under the collectors is in the dark.

Nevada Solar One is an example of a solar trough electricity generation system, with a nominal capacity of 64 megawatts and with a surface area of 320 acres.[26] The plant uses 76 kilometers of troughs that concentrate the sun's rays at focal axes of the troughs. Heat from the fluid is used to generate steam in a steam turbine. The project required an investment of approximately $250 million. Electricity production is estimated to be 136 million kilowatt-hours

24. Kevin Bullis, "Cheaper Solar Thermal Power," July 28, 2009. Available at www.technologyreview.com/news/414511/cheaper-solar-thermal-power/. It is interesting to note that the article mainly concerns converting solar heat to electricity using a Stirling engine rather than the turbines discussed here; however, the proposed prices were still 12 to 15 cents per kWhour, which is much higher than fossil fuels or wind.

25. See www.brightsourceenergy.com/sedc#.U6rRWfldV8E.

26. See www.acciona-energia.com/activity_areas/csp/installations/nevadasolarone/nevada-solar-one.aspx.

per year, which means the average power generated is around 15 megawatts. Thus, the installation cost per average watt is $16.70.

Solar thermal plants in Spain can generate uninterrupted electricity for twenty-four hours straight during the summer.[27] They achieve this by using a molten salt heat storage design. Of course, the total electrical power delivered was less than the electrical power originally generated when the sun was shining, but the storage did allow the energy to be spread over darkness. If conventional fossil fuel sources need to cover the entire load during nonrenewable hours, then the cost of renewables must be increased by the fossil fuel plant cost required to back them when they are not providing power.

The Andasol power plant includes three solar power generators: Andasol 1, 2, and 3.[28] Each generator stores heat in a molten salt mixture of 60 percent sodium nitrate and 40 percent potassium nitrate.[29] A turbine produces electricity using this stored heat during the evening or when the sky is overcast. The 1,010 million kilowatt hours of storage is enough to run a turbine for about 7.5 hours at full load in the absence of sunlight.[30] This process almost doubles the number of operational hours at the solar thermal power plant per year. Like Nevada Solar One, Andasol 1 has a gross electricity output of 50 megawatts, producing 181 million kilowatt hours, corresponding to approximately 21 megawatts of average power.[31] Andasol 1 cost approximately €300 million ($380 million) to build,[32] so the installation cost was around $18 per watt, which is comparable to the cost for Nevada Solar One. Solar millennium, which developed the Andasol power plant, filed for bancruptcy in 2012.[33]

27. See www.torresolenergy.com/TORRESOL/gemasolar-plant/en and www.recharge news.com/solar/article1292212.ece.

28. See www.nrel.gov/csp/solarpaces/project_detail.cfm/projectID=3.

29. See www.nrel.gov/csp/solarpaces/project_detail.cfm/projectID=3.

30. "The parabolic trough power plants Andasol 1 to 3," Solar Millennium. PDF page 13. Available at http://large.stanford.edu/publications/coal/references/docs/And asol1-3engl.pdf.

31. See www.estelasolar.eu/index.php?id=32.

32. David Biello, "How to Use Solar Energy at Night" *Scientific American* February 18, 2009. Available at www.scientificamerican.com/article/how-to-use-solar-energy-at-night/.

33. See www.solarmillennium.de/Technology/References_and_Projects/Andasol __Spain_/The_Construction_of_the_Andasol_Power_Plants_,lang2,109,155.html and www.solarmillennium.de/english/forcreditors/index.html.

Solar thermal systems, like other heat engines, are limited by the Carnot efficiency. In practice, slightly lower results are obtained because of reradiation of heat. Given the large collection areas required and the generating expenses involved, it is unlikely that solar thermal electricity generation will play a significant role in the energy economy, though solar thermal may be useful in particular niche locations that have abundant sunlight and small electricity consumption. In contrast, solar thermal heating for climate control may be able to play a significant role, particularly in warm regions.

Combustion-Based Heating

Fuels that are burned contain stored energy that is released at a chosen time. One way to store excess wind or solar energy would be to create fuels. This is what plants naturally do during photosynthesis. Plants convert energy provided by sunlight into stored chemical energy. That stored chemical energy can be released by burning in an open flame or by "burning" within living beings that eat the plants to obtain energy. Hydrocarbons were once plants that created stored chemical energy from sunlight. They hold that stored chemical energy until enough heat is applied to release the stored energy. If the energy is released steadily in the form of heat, the result is a controlled flame. In the flame, the heat generated by burning releases stored energy at a constant rate, which maintains the flame. In contrast, if the heat generated increases with time, the result is an explosion.[34]

Combustion-Based Systems Details

In all burning-based systems, oxidation releases heat energy because the electrons in the atoms rearrange themselves to lower energy levels, as discussed in online Appendixes 2 and 3. In particular, the negatively charged electrons redistribute themselves so that they are closer to the positive charges in the nuclei than they were in the initial state. The nuclei in these reactions do not change.

---

34. Nuclear reactions are similar: energy is stored in nuclei. In fission, the energy release is triggered by the arrival of a neutron from previous fission reactions. If the energy is released at a constant rate, then controlled fission occurs. If the energy release is sudden, an explosion occurs. In fusion, the energy is stored in nuclei, but like hydrocarbon burning, the energy release is triggered by the heat generated by previous fusion reactions.

Thus, the elements that are present at the beginning of the reaction are the same as the elements that are present at the end of the reaction; however, they may form different molecular combinations.

The chemical expression for hydrogen burning in air is $4H + O_2 \rightarrow 2H_2O +$ Energy. Energy is released because the negatively charged electrons in water are on average closer to the positively charge nuclei toward which they are attracted than they were in the separate hydrogen and oxygen molecules. In hydrogen burning, hydrogen and oxygen simply combine, forming water and no additional products, including no carbon dioxide. Thus, hydrogen burning is the cleanest burning reaction, resulting in no undesirable emissions; however, hydrogen is not readily available.

One way to generate free hydrogen is to use electrical energy to separate water molecules into hydrogen and oxygen. Once the hydrogen is created, it can be stored for burning later. Excess wind or solar energy could be used to provide the required hydrogen from water, but hydrogen gas is a difficult fuel to use because it is so volatile, as the Hindenburg disaster reminds us. There are various ways of storing hydrogen, including the formation of liquid or solid compounds; however, the energy per unit mass and energy per unit volume are significantly lower than the values for gasoline. Worse, the more stable the storage, the more difficult it is to extract the energy from the storage. Considerable research effort is being spent on efforts to improve hydrogen storage; however, absent good storage solutions, hydrogen burning systems are not yet commercially useful even though hydrogen is clean and pollution free—creating only clean water as by-product.

In contrast, hydrogen trapped in hydrocarbons is stored efficiently and safely in readily available fossil fuels.[35] Like hydrogen burning, hydrocarbon burning releases energy as a result of hydrogen pairing with oxygen to form water; therefore, people obtain the majority of their primary energy by burning hydrocarbons.

For example, natural gas consists predominantly of methane. Methane is a molecule containing one carbon atom and four hydrogen atoms. When methane burns in air, it produces water and releases energy, just as hydrogen burning does; however, since methane also contains carbon, methane burning also generates carbon dioxide, as shown in the equation below.

35. Biofuel burning also depends on the favorable combination of hydrogen with oxygen to form water, and of course hydrocarbons were originally biofuels.

$$CH_4 + 2O_2 \rightarrow CO_2 + 2H_2O + Energy$$

The burning reaction is favorable because the electrons in water have a lower energy than the electrons in free oxygen. The change in the energy stored in the carbon bonds is approximately neutral. Thus, the energy released is approximately proportional to the number of water molecules formed. The generation of carbon dioxide is a fundamental feature of the complete burning reaction.[36] The energy cannot be released without it. The only way to lower the amount of carbon dioxide emitted into the atmosphere as a result of the burning is to somehow capture it after it is created.

It is very important to note that the amount of carbon dioxide emitted when burning fossil fuel to generate a given amount of energy depends strongly on the fossil fuel that is burned. Methane is the simplest hydrocarbon. Pentane is a slightly more complex hydrocarbon, which contains five carbon atoms and twelve hydrogen atoms. The chemical equation for the burning of pentane is shown below.

$$C_5H_{12} + 8O_2 \rightarrow 5CO_2 + 6H_2O + Energy$$

The heat energy from hydrocarbon burning comes primarily from binding the free oxygen to the hydrogen that is released from the carbon, so the energy released increases with the ratio of hydrogen to carbon, which is equal to half the ratio of water molecules to carbon dioxide.

In general, the chemical expression for saturated hydrocarbons is C_nH_{2n+2}, which is consistent with the methane and pentane equations above where $n = 1$ for methane and $n = 5$ for pentane. Similarly, the general expression for hydrocarbon combustion is

$$C_nH_{(2n+2)} + [(n+1)/2 + n]O_2 \rightarrow (n+1)H_2O + nCO_2 + Energy.$$

Substituting $n = 1$ for methane and $n = 5$ for pentane gives us the following:

$$C_1H_4 + 2O_2 \rightarrow 2H_2O + CO_2$$
$$C_5H_{(12)} + 8O_2 \rightarrow 6H_2O + 5CO_2.$$

Thus, the ratio of H_2O to CO_2 is $(n+1)/n$. For methane, where $n = 1$, the ratio is two, whereas for pentane it is 6/5, where higher hydrocarbons have values

36. Incomplete burning occurring in the absence of sufficient oxygen can produce carbon monoxide rather than carbon dioxide.

asymptotically approaching one. Since methane generates the least carbon dioxide per water molecule, burning methane produces the lowest carbon dioxide per energy released. Coal contains many higher hydrocarbons; as a result, coal releases almost twice as much carbon dioxide per energy unit as natural gas, which was shown clearly in Figure 10.3. This fundamental improvement in the number of carbon dioxide molecules released per energy consumed implies that shifting from coal to natural gas would greatly reduce carbon emissions.

The United States has been making precisely this shift during the last five years, resulting in a decrease in carbon dioxide emissions to 1992 levels.[37] Furthermore, oil and coal contain additional material besides hydrocarbons, and those materials can result in significant emissions when coal or petroleum are burned, as I discussed in Chapter 10.

Biomass is also becoming more widely used. It represents various different sources, some of which combust directly and some of which are processed to produce a combustible material. For example, wood is a traditional biomass that is burned directly, and ethanol is a traditional biomass by-product that has long been consumed in the form of beverages but is now becoming an increasingly important alternative to petroleum in automotive transport. In addition, other crops, such as switch grass, and nontraditional plant crops, such as algae, are now being tested as possible biofuel sources. Crop research coupled with genetic engineering may provide a wonderful energy source based on switch grass or algae, but no obvious candidate stands out at present.

Nuclear Power

There are no technical issues with generating electricity from nuclear fission. As long as no unplanned events occur, nuclear fission power plants function well. The two major issues with operating a fission power are a catastrophic runaway reaction as a result of an accident and a terrorist attack. Though engineering can reduce these risks, it can never completely remove them. People, then, are faced with a decision about how much risk they are willing to take to generate power from nuclear fission.

37. See www.epa.gov/climatechange/ghgemissions/gases/co2.html. This website shows a graph of carbon dioxide emission vs time where the source for the data is given as www.epa.gov/climatechange/ghgemissions/usinventoryreport.html.

In addition to the risks of operating plants, there is the unsolved problem of nuclear waste disposal. At present, the United States has no permanent storage facility for high-level civilian nuclear waste, despite decades of drama surrounding the possibility of creating such a repository at Yucca Mountain. Thus, despite the technical feasibility of nuclear power, which does not produce carbon dioxide emissions, the world is moving away from fission-powered electricity generation.

In contrast with fission power, there are no commercial fusion power systems. Such systems have been one to two decades away for the last sixty years. There is no scientific consensus on how or whether a fusion system can provide enough sustained density and temperature to achieve controlled fusion that generates the significant steady net energy release required for electricity generation. Even if sustained fusion is scientifically possible, it is not at all clear that electricity generated by fusion would be economically viable. It is quite possible that fusion-based electricity could be the answer to all of our energy needs, but it is also quite possible that fusion will never be a viable source of electrical power. Either way, we can continue to exploit the renewable energy originating in the steady fusion reactions that occur inside the sun, where gravity provides the conditions required to maintain continuous fusion.

Recommended Steps toward a Renewable Future

Research Areas

1. Large- and small-scale energy storage research
2. Studies on optimizing overall wind harvesting over areas more than 100 square kilometers, including effects on weather
3. High-temperature superconductivity
4. Solar PV cell efficiency
5. Better biomass yields
6. Better systems of converting biomass to usable biofuels
7. Studies of whether combinations of geographically distributed renewable energy sources could meet actual rather than average U.S. energy needs without improved storage

Technological and/or Social Changes Not Requiring New Science

1. Convert to electrical cars, preferably powered by renewable energy. *Consequences:* Lower energy use for any electricity source. Emissions depend on electricity source, where renewables are best and coal-burning plants are worst.
2. Use hybrid cars in urban situations where braking and idling are frequent. *Consequences:* Lower energy use and emissions by recycling energy and avoiding use of the internal combustion energy in regimes far from peak efficiency.

3. Exploit computing power to reduce energy use that does not benefit consumers, including training consumers (e.g., driving around looking for parking spaces, sitting still in traffic jams, heating and lighting empty rooms).
   Consequences: Lower energy consumption overall, reducing all negative energy use consequences and making an all–renewable energy economy more feasible.

4. Use computing power to even out electric power consumption, so that producing plants operate continuously near maximum efficiency, including smart meters that allow consumers to reduce their use during periods of peak demand.
   Consequences: Lower energy consumption and reduce cost by allowing plants to operate near peak efficiency.

5. Put solar panels on as many roofs as possible and create PV solar farms in deserts.
   Consequences: Lower commercial electricity generation by around 30 percent, which would reduce emissions. Lower peak commercial demand on hot sunny days.

6. Install large rotor wind turbines on high towers in windy places.
   Consequences: Provides the most cost-effective wind power.

7. Develop long-distance transmission networks to deliver renewable energy from favorable sites and reduce fluctuations by pooling both supply and demand.
   Consequences: If the network is spread wide enough, the network always includes places where the wind is blowing and/or the sun is shining, allowing continuous renewable energy delivery without storage.

Acknowledgments

The many people who have contributed to Wikipedia and Wikimedia Commons provide tremendous resources, for which I am very grateful. This book would not have been possible without the exceptional mentoring provided by Professor George Whitesides. Harris Baseman's heartfelt advocacy for changes in energy policy contributed significantly to my inspiration for this book. Claudia Danilowicz, Renee Sher, and an anonymous reviewer contributed many valuable suggestions. I also appreciate the help and encouragement provided by Harvard University Press. Finally, Alyson Conover read the book several times, even though she is not very fond of science. Her comments have made the book both clearer and more interesting.

Index